Lecture Notes
in Business Information Processing   93

Series Editors

Wil van der Aalst
  Eindhoven Technical University, The Netherlands
John Mylopoulos
  University of Trento, Italy
Michael Rosemann
  Queensland University of Technology, Brisbane, Qld, Australia
Michael J. Shaw
  University of Illinois, Urbana-Champaign, IL, USA
Clemens Szyperski
  Microsoft Research, Redmond, WA, USA

Stanisław Wrycza (Ed.)

Research in Systems Analysis and Design: Models and Methods

4th SIGSAND/PLAIS EuroSymposium 2011
Gdańsk, Poland, September 29, 2011
Revised Selected Papers

 Springer

Volume Editor

Stanisław Wrycza
University of Gdańsk
Department of Business Informatics
ul. Piaskowa 9
81-864 Sopot, Poland
E-mail: swrycza@univ.gda.pl

ISSN 1865-1348   e-ISSN 1865-1356
ISBN 978-3-642-25675-2   ISBN 978-3-642-25676-9 (eBook)
DOI 10.1007/978-3-642-25676-9
Springer Heidelberg Dordrecht London New York

Library of Congress Control Number: 2011941500

ACM Computing Classification (1998): J.1, H.4, D.2

Typesetting: Camera-ready by author, data conversion by Scientific Publishing Services, Chennai, India

Printed on acid-free paper

Springer is part of Springer Science+Business Media (www.springer.com)

Preface

Systems analysis and design (SAND) has been the central field of research and education in the area of management information systems (MIS) or, as it is called more frequently in Europe, business informatics, almost from the time of its origins. SAND continuously attracts the attention of both academia and business. The rapid progress of information and communications technology naturally generates requirements for new generations of SAND methods, techniques and tools. Therefore, international thematic conferences and symposia have become widely accepted forums for the exchange of concepts, solutions and experiences in SAND. In particular, the Association for Information Systems (AIS) is undertaking an initiative toward SAND's international development.

The objective of the EuroSymposium on Systems Analysis and Design is to promote and develop high-quality research on all issues related to SAND. It provides a forum for SAND researchers and practitioners in Europe and beyond to interact, collaborate and develop their field. The EuroSymposia were initiated by Keng Siau as the SIGSAND—Europe Initiative. Previous EuroSymposia were held at:

- University of Galway, Ireland – 2006
- University of Gdansk, Poland – 2007
- University of Marburg, Germany – 2008

The accepted submissions of EuroSymposium 2007 were published in: A. Bajaj, S. Wrycza (eds), Systems Analysis and Design for Advanced Modeling Methods: Best Practices Information Science Reference, IGI Global, Hershey, New York, 2009. After a three-year break, one of the former organizers, the Department of Business Informatics of the University of Gdansk, decided to re-start the EuroSymposium as a joint undertaking of two AIS units—SIGSAND and PLAIS. Therefore, the three organizers of the 4th EuroSymposium on Systems Analysis and Design were:

- SIGSAND – AIS Special Interest Group on Systems Analysis and Design
- PLAIS – The Polish Chapter of AIS
- The Department of Business Informatics of University of Gdansk, Poland

SIGSAND is one of the most active AIS SIGs with a substantial record of contributions to AIS. It provides services such as the annual North American and European SAND Symposia, research and teaching tracks at major IS conferences, a listserv and special issues in journals.

The Polish Chapter of the Association for Information Systems (PLAIS) was established in 2006 as the joint initiative of Claudia Loebbecke, former President of AIS, and Stanislaw Wrycza, University of Gdansk, Poland. PLAIS co-organizes international and domestic IS conferences.

The Department of Business Informatics of the University of Gdansk conducts intensive teaching and research activities. Some of its academic manuals are bestsellers in Poland. The department is also active internationally, organizing various conferences including the 10th European Conference on Information Systems (ECIS 2002) and the 7th International Conference on Perspectives in Business Informatics Research (BIR 2008). The department is a partner of the European Research Center for Information Systems consortium.

EuroSymposium 2011 - the 4th SIGSAND/PLAIS Symposium on Systems Analysis and Design - was held in Gdansk, Poland, on September 29, 2011.

EuroSymposium had an acceptance rate of 45%, with submissions divided into the following three groups:

− Business Processes Modeling
− Integrated Systems Development
− Software Development

The accepted papers reflect the current trends in the systems analysis and design field.

During EuroSymposium 2011, two keynote speeches were given:

− *Vijay Khatri, Operations and Decision Technologies Kelley School of Business, Indiana University, USA*: Developing an Agenda for Research in Systems Analysis and Design at the Intersection of Design Science and Behavioral Science.
− *Jaroslaw Jackowiak, ISV Technology Manager, Academic Initiative Architect IBM Software*: IBM Cloud Concept and Implementation.

I would like to express my thanks to all authors, reviewers, Advisory Board, International Program Committee and Organizing Committee members for offering their support, effort and time. They made possible another successful Systems Analysis and Design EuroSymposium.

September 2011   Stanislaw Wrycza

Organization

General Chair

Stanislaw Wrycza   University of Gdansk, Poland

Organizers

- AIS SIGSAND - Special Interest Group on Systems Analysis and Design of Association for Information Systems
- PLAIS - Polish Chapter of Association for Information Systems
- Department of Business Informatics, Faculty of Management of University of Gdansk

Advisory Board

David Avison	ESSEC Business School, France
Richard Baskerville	Georgia State University, USA
Sjaak Brinkkemper	Utrecht University, The Netherlands
Gordon B. Davis	University of Minnesota, USA
Phillip Ein-Dor	Tel Aviv University, Israel
Guy Fitzgerald	Brunel University Uxbridge, UK
Joey F. George	Florida State University, USA
Dimitris Karagiannis	University of Vienna, Austria
Julie E. Kendall	Rutgers University, USA
Claudia Loebbecke	University of Cologne, Germany
John Mylopoulos	University of Toronto, Canada
Keng Siau	University of Nebraska-Lincoln, USA

International Program Committee

Stanislaw Wrycza	University of Gdansk, Poland (Chair)
Eduard Babkin	Higher School of Economics, Moscow, Russia
Akhilesh Bajaj	University of Tulsa, USA
Dinesh Batra	Florida International University, Miami, USA
Jan vom Brocke	University of Lichtenstein
Glenn J. Browne	University of Virginia, Charlottesville, USA
Rimantas Butleris	Kaunas University of Technology, Lithuania
Sven Carlsson	Lund University, Sweden
Marco De Marco	Sacro Cuore Catholic University of Milan, Italy
Bjoern Erik Munkvold	University of Agder, Norway
Andrew Gemino	Simon Fraser University, Burnaby, Canada

Rolf Granow	Luebeck University of Applied Sciences, Germany
Alan R. Hevner	University of South Florida, Tampa, USA
Seamas Kelly	University College Dublin, Ireland
Vijay Khatri	Indiana University Bloomington, USA
Marite Kirikova	Riga Technical University, Latvia
Andrzej Kobylinski	Warsaw School of Economics, Poland
Karl Kurbel	European University Viadrina Frankfurt (Oder), Germany
Miroslawa Lasek	University of Warsaw, Poland
Bogdan Lent	University of Applied Sciences, Zurich, Switzerland
Leszek Maciaszek	Macquarie University Sydney, Australia and Wroclaw University of Economics, Poland
Bartosz Marcinkowski	University of Gdansk, Poland
Jacek Maslankowski	University of Gdansk, Poland
Nava Pliskin	Ben-Gurion University of the Negev, Israel
Paul Ralph	Lancaster University, UK
Michael Rosemann	Queensland University of Technology, Australia
Matti Rossi	Aalto University School of Economics, Finland
Reima Suomi	University of Turku, Finland
Carson Woo	University of British Columbia, Canada
Iryna Zolotaryova	Kharkiv National University of Economics, Ukraine
Joze Zupancic	University of Maribor, Slovenia

Organizing Committee

Stanislaw Wrycza, Anna Szynaka, Lukasz Malon, Bartosz Marcinkowski, Jacek Maslankowski
Department of Business Informatics, University of Gdansk, Poland

Table of Contents

Part I

Business Processes Modeling

Reengineering University: Modeling Business Processes to Achieve Strategic Goals

Aleksey Shutov

25/12, Bolshaya Pecherskaya str, Nizhny Novgorod, Russia, 603155
ashutov@hse.ru

Abstract. Recognizing the challenges that a contemporary university faces, an inter-faculty group of researchers performed the analysis of organizational and management approach in one of Russia's universities and modeled existing business processes. The university is a multi-campus organization and to manage it efficiently it is necessary to make transition from the traditional management model to a more flexible - network structure. This paper focuses on development of the Goal Tree and designing of AS-IS model. The researchers applied IBM Business Modeler [1] which allows to execute modeling and simulation of business processes. The practical value of the proposed approach is that it enables the university management to improve performance and effectiveness.

Keywords: Network university, business process modeling, organizational structure, simulation modeling.

1 Introduction

The paper examines the challenging tasks of analyzing and assessing University organizational and management model, and the role of business processes in achieving the strategic goals of the University. The assessment will be based on data obtained from simulation modeling and on the developed methods of assessment business processes-on-goals impact. The business processes are modeled with specialized software and described with BPMN (Business Process Modeling Notation) [2] [3]. The correlation between processes and goals is defined with the help of the developed Goal Tree.

A modern university must conform to the needs of society and respond quickly to social changes. Not only should universities prepare specialists, they must become centers of scientific knowledge, academic research and development. Furthermore, inter-university collaboration encouraged by the Bologna process is strengthening: distributed universities emerge which represent international academic communities which use the same curriculum and support student and teacher mobility. Russian universities actively participate in this process [4].

This paper considers National Research University Higher School of Economics (HSE) as the example and source of initial data. Until 2010 this institution was called State University Higher School of Economics and included several campuses: Nizhny Novgorod, St.-Petersburg, Moscow, Perm'. Each campus had a relatively independent

S. Wrycza (Ed.): SIGSAND/PLAIS 2011, LNBIP 93, pp. 3–14, 2011.

organizational structure. In 2011 HSE was awarded the status of National Research University (NRU HSE) which lead to reengineering and unification of business processes in all branches, and transition from mechanistic, closed and function-oriented model to open and flexible conception promoting management based on self-organization, constant growth and adaptation.

Accomplishment of this task requires new methodology to define university management and organizational on model. It will consider the university as a specific type of organization where a part of stakeholders – students - participate in business processes.

The methodology which we aim to develop will allow the administration of the university to strategize and manage the development based on principles of self-organization and open service-oriented approach to identifying mission, purpose and goals of the university. This opportunity will be realized on the basis of developed models which address social, economic, legal, cultural and technological issues and take into account limitations of university functioning.

Our research consists of the following stages:

1. Study the subject area, identify strategy-based business goals of the university.
2. Create a Goal Tree for the university and develop the matrix of correlation between university processes and established goals, define methods of calculation of process-on-goal impact.
3. Develop an AS-IS model, check the model's adequacy, analyze the model, etc.
4. Develop a TO-BE model, analyze the model.
5. Calculate weighing coefficients for university business processes taking into account their process-on-goal impact.
6. Apply the developed methodology to university development process.

The presented research opens the series of articles devoted to optimization of conceptions of design and implementation of management model for contemporary university. The article focuses on development of AS-IS model. The interdisciplinary scientific approach based on the latest international research has been used to achieve these goals.

The following tasks were completed within this research:

- Goal Tree and Matrix of correlation between university processes and established goals developed, methods of calculation of processes-on-goal impact are defined.
- Existing key business processes of university are analyzed and described (admission procedure, teaching, organization of academic activity, financing of main processes within the university).

The content of the article was divided into the following primary parts:

- Part 2 reviews the literature on organization structure modeling.
- Parts 3 and 4 focus onGoal Tree development, algorithms used to calculate relationship between goals and processes, design of the model.
- Part 5 presents the obtained results.

2 Literature Review

It has been mentioned above that to achieve growth and to occupy leading positions in education and science it is essential for the University to adjust its management approach and provided services to new conditions. Solving these problems requires changes in university operation. These processes need full-scale, thorough study and analysis of all potential development options, taking into account process-on-goal impact.

Modeling of organization processes has been studied by researches and scientists from such different areas of knowledge as sociology, mathematics, cybernetics and others since 1970s. The researches mainly focused on proving the effectiveness of network organizations basing on the results of social surveys and mathematical calculations. The authors explained failures of some network organizations, modeled their operations and studied the impact of human factor on the functioning of an organizations. M. Granovetter [5-7] explains that an individual makes decisions depending on his benefits and costs and on what others are doing. Eng Teck-Yong [8] shows how the Internet enhances effectiveness of network organizations. The study about Australia's network universities [9] describes the reasons why federated universities could not use their network to achieve a competitive advantage. Other possible causes of failure in network organizations are presented in *California Management Review* [10]. In [11] the authors made an attempt to describe the model of learning process in a network organization basing on their six years studying of a conglomerate. In [12] the authors apply the notion of network organization to universities. The article analyses operations of five Australia's universities, namely influence of new technologies on higher education, and advantages and disadvantages they provide to academic activity. These works are based on experiments and sociological surveys. In [13] the researcher describes changes in network organizations and makes an attempt to identify factors influencing changes in organizational structure. For the experiment were chosen a military group and a network organization, and the tools of dynamic analysis (DNA, SNA) were applied. Russian Center of studies LANIT describes the main factors that are necessary for universities to succeed in conditions of diversification of state financial support to science and research [14]. In [15], basing on the main principles of network economy, the following strategic development axes in network education can be identified: develop and standardize software used in network education, transform the structure of university, give more independence to faculties and departments, diversify educational services, develop customized education, increase communication and collaboration among various educational institutions and etc.

Other similar publications have also been studied, but the controlling idea is obvious: all these works promote network organizations as an effective structure. Some articles contain descriptions of mathematical models of organization and statistical methods, however the simulation modeling is rarely applied. Simulation modeling and modeling of business processes (its subset) can be helpful when studying the behavior of complex objects. In this paper the task to analyze organization's behavior is approached using modeling of business processes because its tools allow to simulate, analyze and optimize business-models, also applying a wide range of mathematical methods to describe behavior and relationship of objects.

3 Work Description and Development of the Goal Tree

In this section we describe how the data for building a model was collected and how we developed the Goal Tree.

Research Description
The works consisted of several stages and involved specialists from the Faculty of Business Informatics and the Faculty of Management. The researchers used iterative approach, i.e. the task was divided into several stages and at the completion of each stage the achieved result was analyzed and, if necessary, refined or specified. Here are the main stages of work: develop questionnaires for university staff, question the staff, develop the model, analyze results, refine and specify processes, revise the model. Some stages were repeated until the required result was achieved. The stages are expanded on below.

Develop Questionnaire
The questionnaire was aimed at obtaining the following data: goals of university and ways of achieving them, business processes of university, inter-department collaboration to ensure the processes implementation, qualitative problems of processes.

The developed questionnaire consists of three sections: general questions (structure, staff, general information) goals of university, description of university processes. The third and largest section of the questionnaire includes questions about: name of the stage, department in charge, list of tasks executed at each stage, description of tasks, relationship between tasks, input/output documentation, etc.

Questioning
The following people responsible for main processes in the organization were surveyed: Director, Head of Studies, Financial Advisor, Heads of Departments. As the result we could study regulatory documents and describe admission and teaching procedures, management of academic process. A completely fulfilled questionnaire thereby gives a full picture of the processes inside organization and provides the base for building the Goal Tree.

Development of the Goal Tree
All processes of the organization must directly or indirectly ensure that business goals of the organization are achieved, the latter are established in the programme of development. To identify the process-on-goal impact, the Goal Tree was built and the coherence between goals and processes was defined.

The main goals of the University (according to the programme of development) are [16] to become a leader in educational market, develop innovation competencies and research, address social and economic issues.

The following major business processes were identified during the research: admission of students, provision of educational services to students (both state-funded and those on a paying basis), courseware for academic process, financial flows, scientific research in the university.

The results are presented in a table (matrix of coherence). In Table 1 is presented a part of the matrix describing the strategic goal – «Becoming a leader in educational market».

Table 1. Matrix of coherence

Becoming a leader in educational market		Admission of fee-paying students (v1)	Admission of state-funded students (v2)	State-funded students (v3)	Fee-paying students (v4)	v5	Community charges (v6)	Function of sub-goals	Superposition of sub-goals	Function of goal
	1. Develop efficient system of staff training								$FA1=k1*A1+K2*A2$	
	1.1 Develop new structure of teaching			$a11$	$a12$	$a13$		$A1=a11*v3+a12*v4+a13*v5$		
	1.2 Attract talented and graduate post-graduate students	$a21$	$a22$	$a23$	$a24$			$A2=a21*v1+a22*v2+a23*v3+a24*v4$		
	2. Develop int. Masters programmes	$b11$	$b12$	$b13$	$b14$	$b15$		$A3=b11*v1+b12*v2+b13*v3+b14*v4+b15*v5$	$FA3=A3$	$F=m1*FA1+m2*FA3+m3*FC3$

Rows represent goals and sub-goals of the university, columns describe the main processes. Cells at intersections of goals and processes contain coefficients that show the impact of a process on achieving the goal. The value corresponding to each process $(v_i,)$ is the assessment obtained from simulation modeling.

It is suggested that the impact of process' execution on goals be calculated as follows:

- Simulation modeling is executed.
- As the result, for each process value v_i is calculated, which can be time of execution, cost, income, delays, etc.
- Basing on the matrix of coherence, values of functions for sub-goals are calculated. For example, $A1=a11*v3+a12*v4+a13*v5$ –value which shows how the processes influence the sub-goal, e.g. «Create new structure of academic process», where $a_{i,j}$ is the weighing coefficient of the process's impact on achievement of this goal.
- Values of functions for goals are calculated, taking into account the already obtained results (Superposition of Goals).
- Final value of processes' impact on a strategic goal is calculated. («Strategic goal»)

In general, calculation of impact of a process on a strategic goal is made with a linear function such as (1):

$$f = k_1 p_1 + k_2 p_2 + ...+ k_n p_n \qquad (1)$$

where f – goal, k_i – weight of an i- process in achievement of the goal, p_i - result of the process' execution, n – number of processes. Values of k_i and the choice of result of p_i execution will be defined further in the research.

4   Designing the Model of University Business Processes

This section describes the University business processes. The model uses BPMN (Business Process Modeling Notation) [2] and description is based on IBM WebSphere Business Modeler[1] [1] [3].

Admission of applicants. All applicants fill in electronic or paper form and may immediately conclude the contract for provision of educational services and pay 25% of the required fee. Once the acceptance of forms is finished, the applicants are admitted to state-funded places basing on USE (Unified State Examination) marks. If an applicant concluded the contact but has a USE mark high enough to be admitted on a state-funded basis, the paid sum is reimbursed. The applicants not admitted on a

---

[1] IBM WebSphere Business Modeler is used within the programme «IBM Academic Initiative ».

state-funded basis are offered admission on a fee-paying basis. By the beginning of the semester new fee-paying students must pay 25% of the tuition fee or provide a letter of guarantee.

The details of the process are shown in fig. 1.

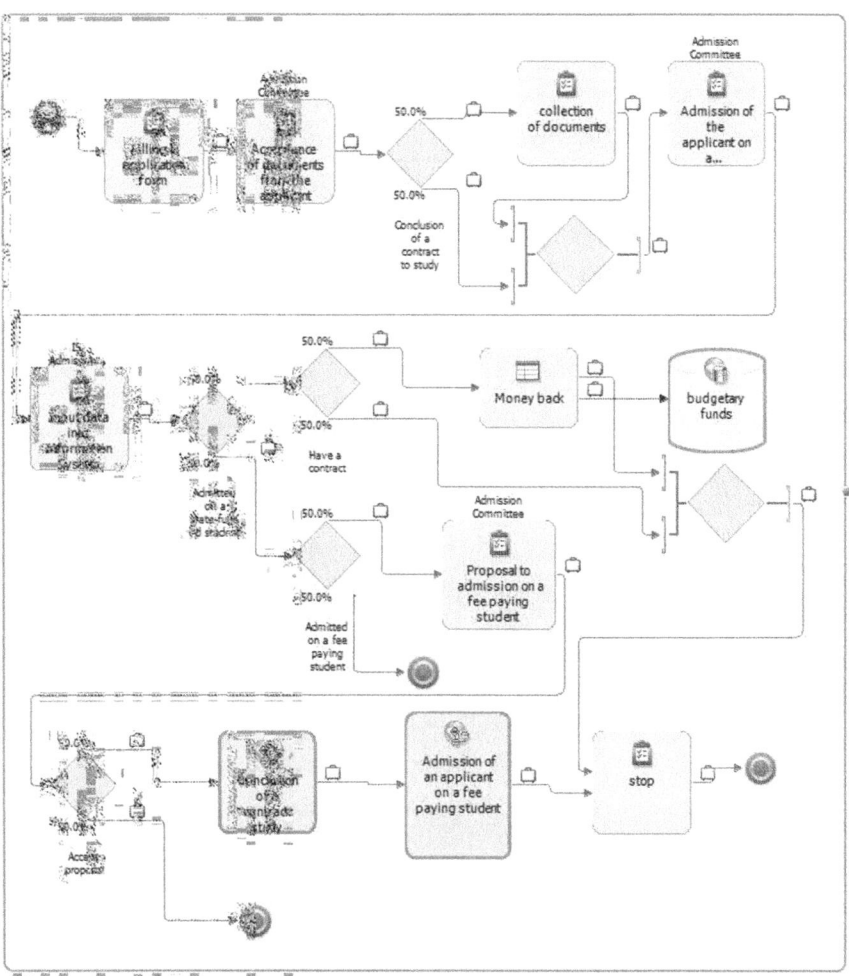

Fig. 1. Admission of applicants

Provision of educational services. This process can be divided into two main processes – teaching and courseware (fig. 2). To make calculations easier, teaching of state-funded and fee-paying students was considered separately and further the process was divided by faculties.

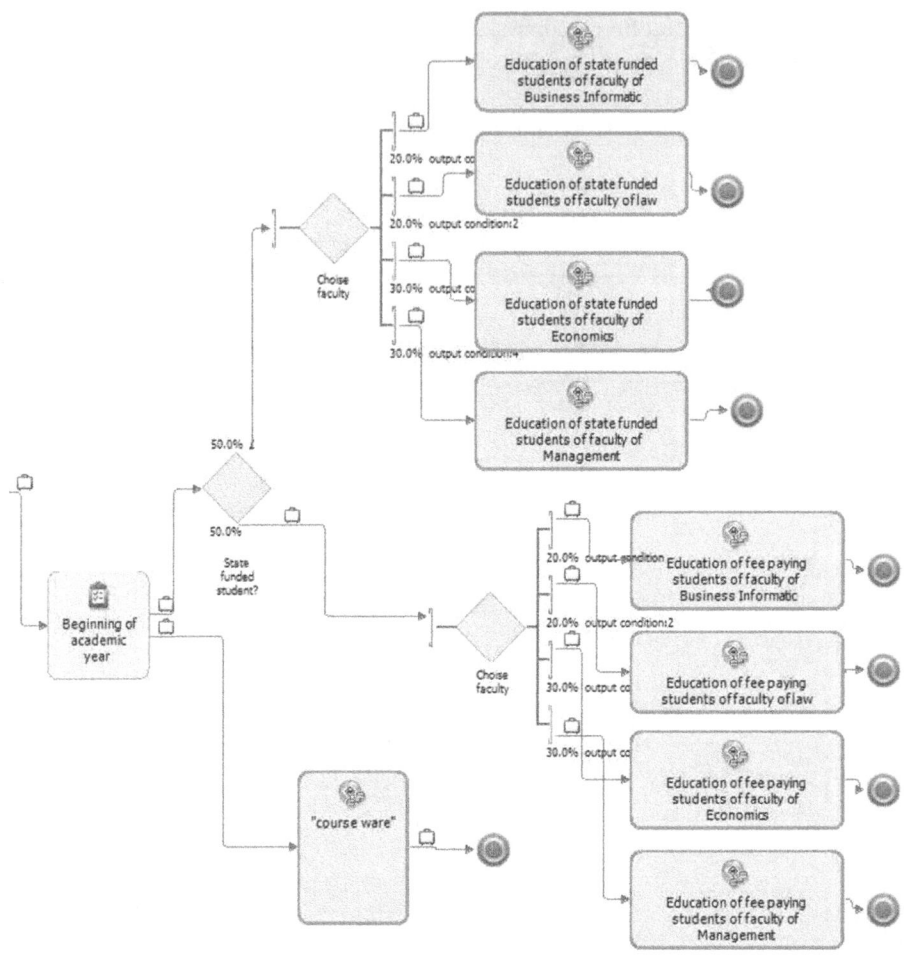

Fig. 2. Provision of educational services

Provision of educational services to state-funded students. After admission the university delivers academic services to its students. During studying the grades of a student are checked and, should they be unsatisfactory, the student is expelled from university. After the first year 5-10% of students are expelled. The process for fee-paying students has the same algorithm but more tasks.

Courseware. Courseware is preparation of teaching loads and programmes. The process of courseware preparation for each academic year is initiated by the main campus in Moscow four or five months in advance and consists of several stages shown in fig. 3. This process is to be accomplished by the beginning of the academic year.

Composition of each type of curricula includes drafting and agreement with the main campus and may be done in several rounds.

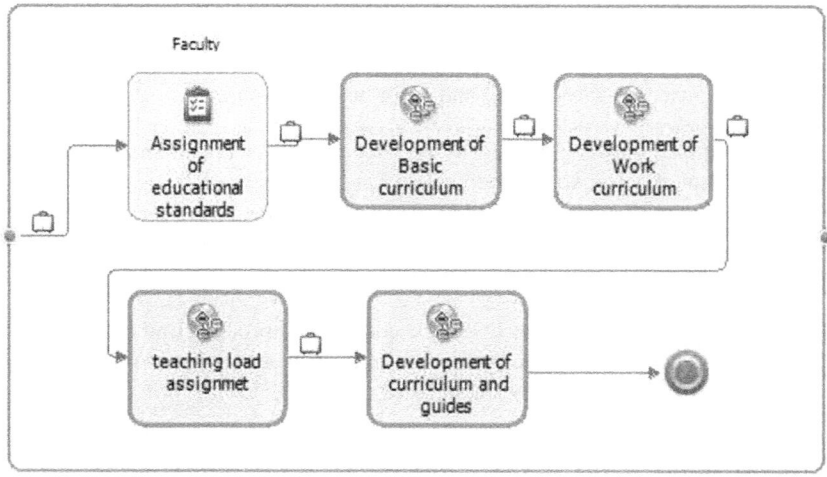

Fig. 3. Courseware

Receipt and distribution of funds. Fig. 4 shows the main budget items such as income and expenses.

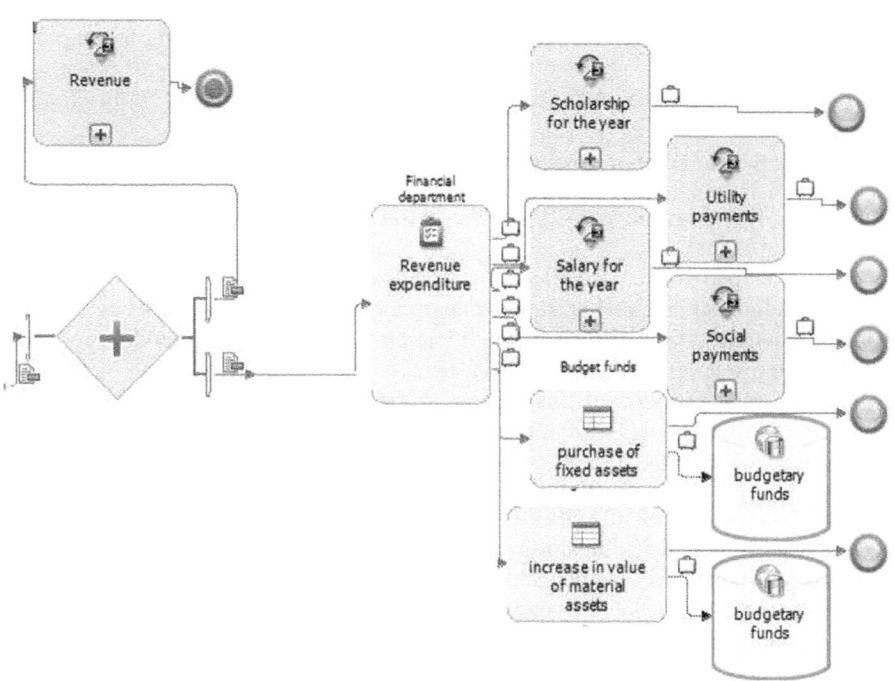

Fig. 4. Receipt and distribution of funds

The University is a state-funded institution, all its financial resources are controlled by the state and all income and expenses are supervised by the exchequer. The budget is planned annually and has quarterly layout.

The funds consist of state-funded and extra-budgetary parts:

- The state-funded part is allocated by the Department of Presidential Affairs and is based on the number of state-funded places at the university.
- Extra-budgetary part – money earned by the university (contracts, fee-paying students, etc). This part is calculated basing on the plan of admission of fee-paying students.

This is one of the main activities to ensure educational process, and at the same time high-quality education is the means of attracting new students therefore increasing the extra-budgetary part of university financial resources.

5   Conclusion

A comprehensive reengineering of university processes is a complex and sophisticated task. We would like to emphasize that this paper presents only the commencement of the research. Within the first stage we have obtained the following expected results:

- Describing existing processes at the example of HSE.
- Building the Goal Tree and defining the methodology of calculating the impact of processes on business goals. (Exact values of weighing coefficients of process-on-goal impact and values obtained from modeling will be calculated further for TO-BE model).
- Building the model of processes.

The Goal Tree was developed on the basis of the programme of University development [16].

The developed model considers the organizational structure of the university, including administrative and academic departments; staff are represented as a pool of human resources with different qualification. Moreover, the model considers information resources, such as various systems.

The researchers conducted simulations for the periods of one, two and three years on the basis of the developed model. The obtained output on the main processes corresponds to real data. The processes were verified by the following parameters:

- Input data: number of students at the beginning of simulation, number of applicants at the beginning of simulation, cost of man-hour for different categories of resources, number of staff in each department.
- Output data: financial expenses, income, temporary costs of execution of processes, labor input for processes, work load.

The work shows the potential and applicability to use the model description (in BPMN) and simulation modeling for various types of analysis and to apply this method to perform such complex tasks as describing and modeling of operation of the whole organization and down to its separate departments.

The outcomes presented above will serve as basis for further research in the field of university reengineering. We aim to tackle the following problems:

- Define coefficients of process-on-goal impact.
- Build a «TO-BE» model for NRU HSE as a network university, including all campuses.
- Analyze the results of modeling a distributed university (using Goal Tree) and prepare recommendations for university strategizing and managing.

References

1. IBM. WebSphere Business Modeler Advanced. IBM Corporation portal,
   http://www-01.ibm.com/software/integration/
   wbimodeler/advanced/
2. OMG. Documents Associated with Business Process Model and Notation (BPMN) Version 2.0. Object Management Group/Business Process Management Initiative (January 2011), http://www.bpmn.org/
3. Wahli, U., et al.: Business Process Management: Modeling through Monitoring Using WebSphere V6.0.2 Products. IBM Corporation (August 2007),
   http://www.redbooks.ibm.com/redbooks/pdfs/sg247148.pdf
4. Organizatsia setevogo vzaimodeystvia vuzov-uchastnikov Bolonskogo processa kak osnova upravlenia integratsiey rossiyskoi sistemy vyshego professionalnogo obrazovania v obsheevropeyskuju: praktika organizatsii, tseli, funktsii, struktura, perspektivy. Universitetskoe upravlenie 1(41), 58–70 (2006)
5. Granovetter, M.: The Strength of Weak Ties. American Journal of Sociology 78, 1360–1380 (1973)
6. Granovetter, M.: Threshold Models of Collective Behavior. The American Journal of Sociology 83(6), 1420–1443 (1978)
7. Granovetter, M., Soong, R.: Threshold models of interpersonal effects in consumer demand. Journal of Economic Behavior & Organization 7, 83–99 (1986)
8. Eng, T.-Y.: An investigation Internet coordination mechanism in network organizations. Journal of Interactive Marketin 21(4), 61–75
9. Massingham, P.: Australia's Federated Network Universities: what happened? Journal of Higher Education Policy and Management 23(1), 19–32 (2001)
10. Raymond, M.E., Snow, C.C.: Causes of Failure in Network Organizations. California Management Review 34(4), 53–72 (1992)
11. Hanssen-Bauer, J., Snow, C.C.: Responding to Hypercompetition: The Structure and Processes of a Regional Learning Network Organization. Organization Science 7(4), 413–427 (1996)
12. Lewis, T., Marginson, S., Snyder, I.: The Network University? Technology, Culture and Organisational Complexity in Contemporary Higher Education. Higher Education Quarterly 59(1), 56–75 (2005)
13. Graham, J.M.: Dynamic network analysis-based communication network evolution and shared situation awareness estimation in the network organization. Carnegie Mellon University, Ph.D. 127 pages. AAT 3171947 (2005)

14. NOU UC "Setevaya Akademia LANIT". Analiz mejdunarodnogo opyta strategicheskogo upravlenia, organizatsii i razvitia universitetov, integrirujuschih peredovye nauchnye issledovania i obrazovatelnye programmy, reshauschih kadrovye i issledovatelskie zadachi obshenatsionalnyh proektov i zadachi regionov. UC Setevaya Akademia, http://univer.academy.ru/Documents/Анализмеждународногоопыта.pdf
15. Ignatyev, A., Portnova, N.: Setevaya ekonomika i obrazovanie. Nizhny Novgorod: Izdatelstvo NNGU, Vestnik Nizhegorodskogo universiteta im. N.I. Lobachevskogo, seria Ekonomika i Finansy, pp. 358–367 (2005)
16. Strategia Nizhegorodskogo filiala Gosudarstvennogo universiteta-Vyshaya Shkola Ekonomiki do 2020 goda (2010)

SysML Requirement Diagrams: Banking Transactional Platform Case Study

Stanislaw Wrycza and Bartosz Marcinkowski

University of Gdansk, Department of Business Informatics, Piaskowa 9,
81-864 Sopot, Poland
{stanislaw.wrycza,bartosz.marcinkowski}@ug.edu.pl

Abstract. The aim of this paper is an investigation into the practicality of SysML Requirement Diagrams in requirement identification and specification for an Internet banking transactional platform. The research attempts to discover if SysML Requirement Diagrams can be applied not only for defining requirements in engineering domains, but also for conducting business cases. The paper is structured as follows: after the Introduction, the prerequisites of requirement specification in SysML are clarified in Section 2. Section 3 exemplifies the containment relationships in a banking transactional platform case study. A Requirement Diagram of a Money Transfer Service in both graphical and tabular form is developed and explained in Section 4, followed by Conclusions. The study has generated demand for more in-depth analysis of business requirements specification using other functionally related SysML/UML diagrams.

Keywords: System Requirements Specification, SysML, Requirement Diagram, UML 2.x.

1 Introduction

Unified Modeling Language (UML) is the most popular modeling language used in software engineering [15]. Numerous examples of its practical applications may be found in literature – [1], [5], [6], [10], [12], [23] to mention just a few. UML-based projects have inspired experts from other fields of engineering, such as process, mechanical, electrical or chemical engineering.

As a result, a technical-targeted profile of UML, named Systems Modeling Language (SysML), has been proposed [14]. This includes nine types of diagrams that have the following sources:

- diagrams specific for SysML only: Requirement Diagrams and Parametric Diagrams;
- modified and extended UML 2.x diagrams: Block Definition Diagrams, Internal Block Diagrams and Activity Diagrams;
- diagrams directly transferred from UML 2.x: Use Case Diagrams, State Machine Diagrams, Sequence Diagrams and Package Diagrams.

Unlike UML 2.x, the diagrams in SysML are organized within three, as opposed to two, generic groups: Behavior Diagrams, Structure Diagrams and Requirement

S. Wrycza (Ed.): SIGSAND/PLAIS 2011, LNBIP 93, pp. 15–22, 2011.

Diagrams. The latter consists of a single type of diagram, i.e. a Requirement Diagram. It is the Requirement Diagram that attracts the most attention from researchers. There are two reasons for this:

- its novelty: this type of diagram did not exist earlier in UML;
- the assessment of the practicality of the Requirement Diagram in practical engineering and business applications.

The above prerequisites have motivated authors to take up the research in the banking industry, research which is currently in progress. Therefore, the aim of the paper is an investigation into the practicality of SysML Requirement Diagrams in requirement identification and specification for an Internet banking transactional platform.

2 SysML for Requirements Specification

The requirements to be generated in the case study of a banking transactional platform may be divided into:

- requirements already accepted by client: functional requirements and non-functional requirements;
- requirements under negotiation.

Differentiation of functional and non-functional requirements is a common business practice. Underlying theoretical concepts are given and clarified in more detail in [11].

As a support in specifying banking transactional platform requirements, a repository of system requirements specified in other projects was used. This classification is presented in Fig. 1 using UML/SysML Package Diagram notation.

Investment Fund Service is the requirement selected for this case study from the collection of *Functional Accepted Requirements* in the *System Requirements Repository* of Banking Transactional Platform requirements. The choice was dictated by the criteria of functional complexity of the requirement and its significance for bank clients.

The basic tool of requirements specification in SysML language is the Requirement Diagram. It has two essential terminological and graphical notions: Requirement and Relationship. In turn, Relationships can be classified into Containment, as well as six types of Dependencies:

- «deriveReqt»,
- «satisfy»,
- «copy»,
- «verify»,
- «refine»,
- «trace».

As the current paper assumes basic knowledge of SysML, these notions will not be expanded on.

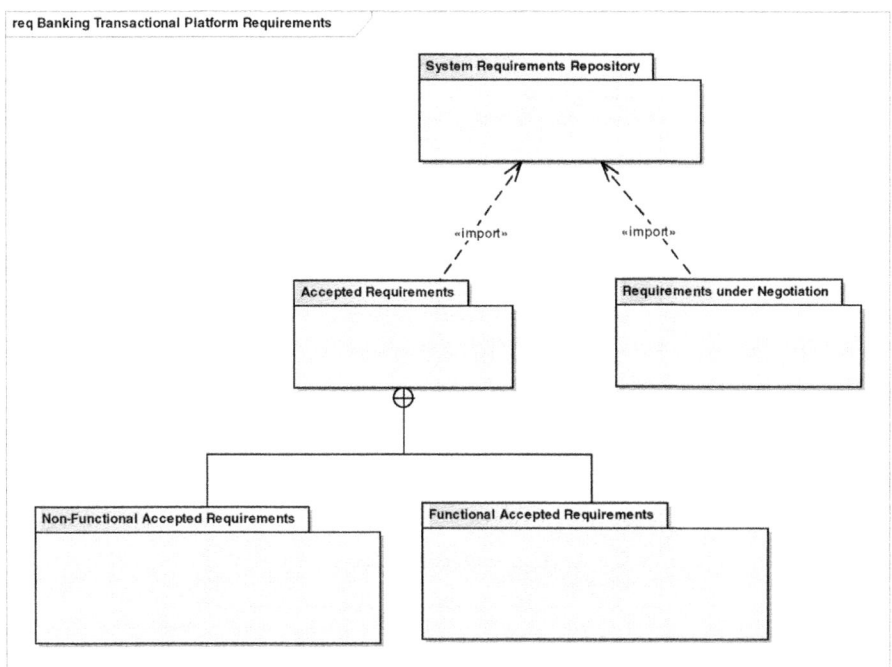

Fig. 1. Packages of requirements of Banking Transactional Platform

The node of requirement contains its name and several sections. The semantics of requirement description can be rich and may include: id, text, priority, obligation, type, risks as well as other optional characteristics, according to developers' decisions and demand for precision in the requirement description. The node of requirement in the hierarchy of requirements which forms the Requirement Diagram is specified by at least the requirement name, its id and text. It is common practice, however, for developers to hide detailed properties in order to reduce the complexity of the specific diagram. In this case, when the specific requirement description is extensive, the alternative, more compact forms of the requirement notion can be used, as shown in Fig. 2. Furthermore, individual requirements may be stereotyped graphically.

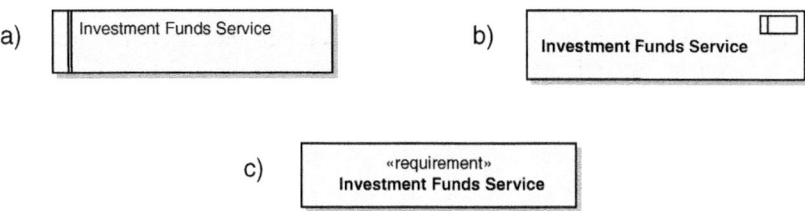

Fig. 2. Compact graphical notations of requirement

Detailed description of Investment Funds Service requirement is illustrated in Fig. 3.

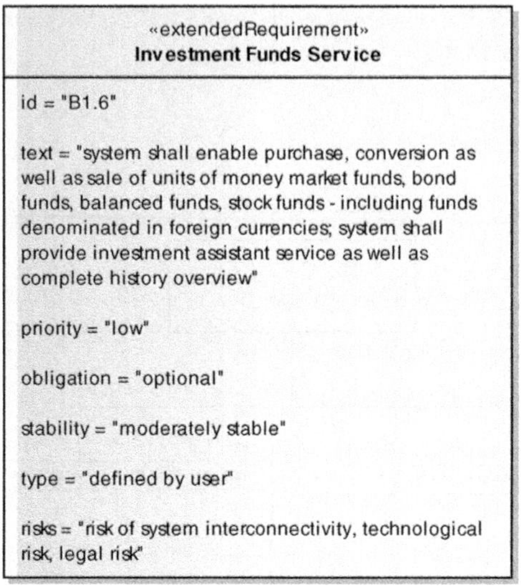

```
                «extendedRequirement»
                Investment Funds Service

id = "B1.6"

text = "system shall enable purchase, conversion as
well as sale of units of money market funds, bond
funds, balanced funds, stock funds - including funds
denominated in foreign currencies; system shall
provide investment assistant service as well as
complete history overview"

priority = "low"

obligation = "optional"

stability = "moderately stable"

type = "defined by user"

risks = "risk of system interconnectivity, technological
risk, legal risk"
```

Fig. 3. Properties of Investment Funds Service requirement

3   Containment Relationships in Banking Transactional Platform Case Study

The hierarchy of sub-requirements in the *Investments Funds Service* requirement is described by the notation of SysML requirements and its containment relationships, as presented in Fig. 4, which includes the following requirements:

- *Current Payments Service,*
- *Credit Cards Service,*
- *Bank Deposits Service,*
- *Debts Service,*
- *Personal Insurances Service,*
- *Investment Funds Service,*
- *Stock Exchange Service.*

Each of the requirements listed may be decomposed into successive requirements, the *Current Payments Service* requirement, for instance, consisting of the following sub-requirements:

- *Bank Accounts Service,*
- *Money Transfers Service,*
- *Fixed Orders Service.*

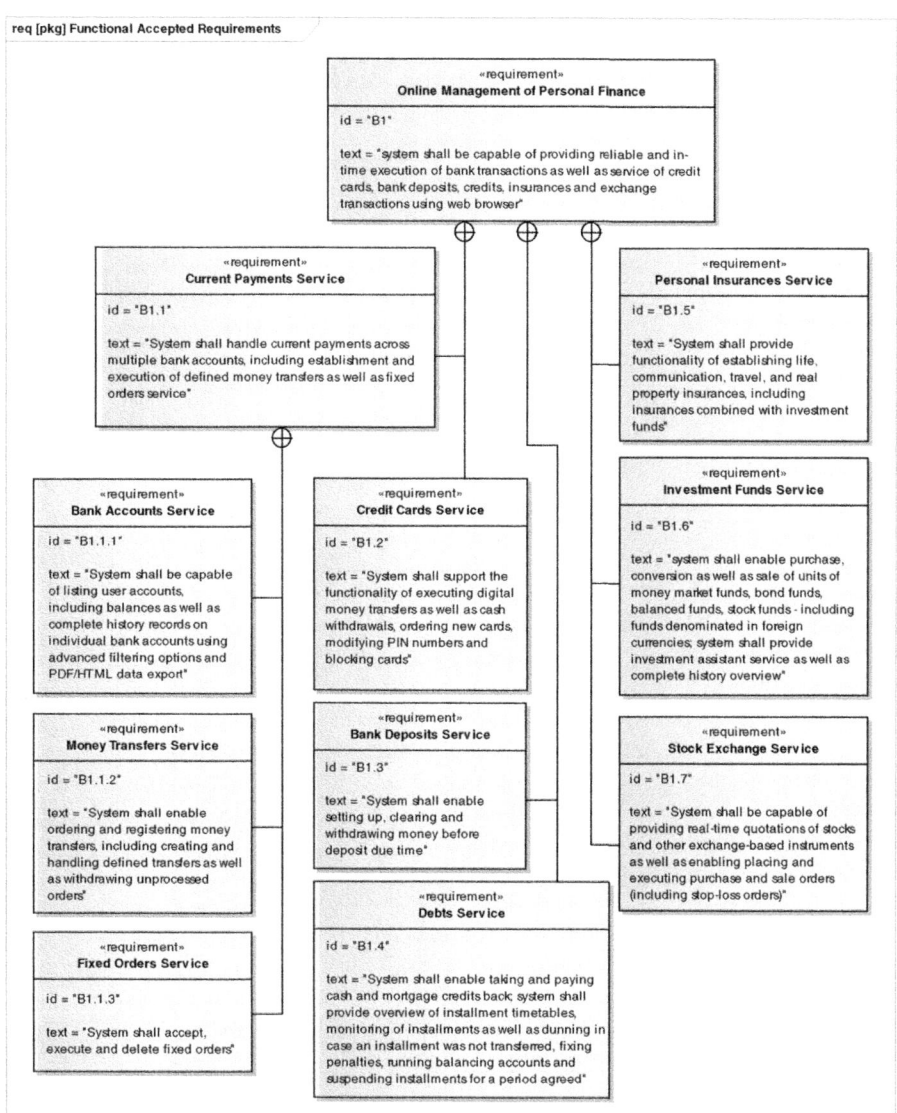

Fig. 4. Hierarchy of Accepted Functional Requirements

4 Requirement Diagrams Elaborated

The complete SysML Requirement Diagram, resulting from the *Money Transfers Service* requirement is presented in Fig. 5. Each of the requirement nodes is described by name only. This requirement diagram includes a number of relationships such as containments and «trace», «copy» and «deriveReqt» dependencies.

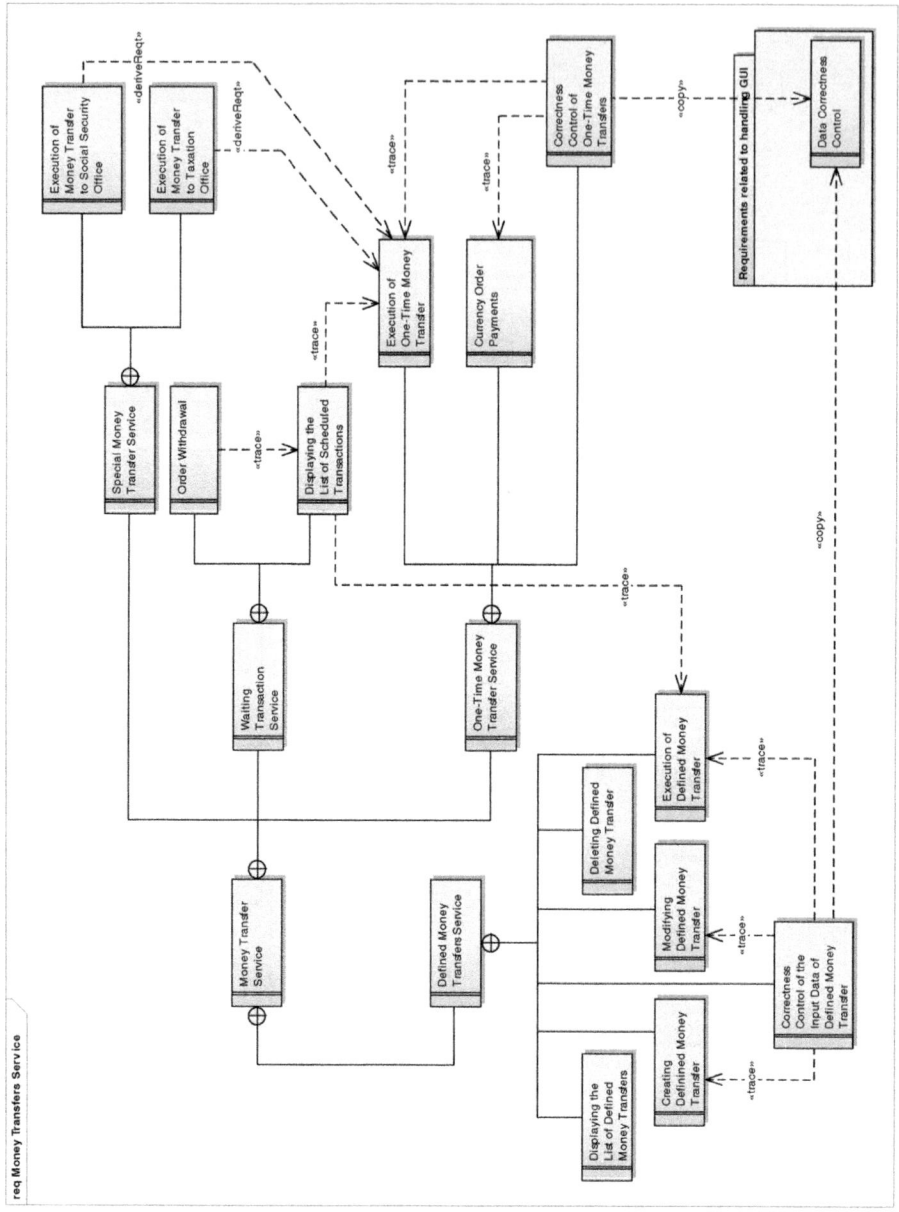

Fig. 5. Decomposition of Money Transfers Service requirement

The relationships used in the Requirement Diagram for *Money Transfer Service* may be presented in tabular form. This allows the relationships (containments or dependencies) between target and source requirements, as shown in Table 1, to be specified.

Table 1. Tabular notation of relationships among requirements

Target requirement		Relationship	Source requirement	
ID	Name	type	ID	Name
B1.1.2.1.5	Execution of Defined Money Transfer	containment	B1.1.2.1	Defined Money Transfers Service
B1.1.2.2.1	Execution of One-Time Money Transfer	containment	B1.1.2.2	One-Time Money Transfers Service
B1.1.2.3.1	Displaying the List of Scheduled Transactions	«trace»	B1.1.2.1.5	Execution of Defined Money Transfer
B1.1.2.3.1	Displaying the List of Scheduled Transactions	«trace»	B1.1.2.2.1	Execution of One-Time Money Transfer
B1.1.2.3.2	Order Withdrawal	«trace»	B1.1.2.3.1	Displaying the List of Scheduled Transactions
B1.1.2.2.2	Execution of Money Transfer to Social Security Office	«deriveReqt»	B1.1.2.2.1	Execution of One-Time Money Transfer
B1.1.2.2.3	Execution of Money Transfer to Taxation Office	«deriveReqt»	B1.1.2.2.1	Execution of One-Time Money Transfer

5   Conclusions

In the current paper an investigation into the suitability of SysML Requirement Diagrams in identifying banking transactional platform requirements was conducted. The research and experiments allow the following conclusions to be drawn:

- SysML Requirement Diagrams are a useful tool and may be applied not only for defining requirements in the engineering domain, but also for conducting business cases;
- SysML Requirement Diagrams facilitate correct specification of the business requirements, their nodes and the versatile relationships – containments and dependencies – among them; they are, however, not sufficient for precise definition of requirement functionality – this aim can be achieved with other SysML/UML diagrams;
- the process of requirement specification is acceptable and easily comprehensible to users cooperating with system designers; the notation, applied in Requirement Diagrams, may be described as user-friendly;
- requirement models of modern systems are complex and must be structured with Package Diagrams;
- Requirement Diagrams are compatible with the stage of requirement specification in object-oriented methodologies of information systems development, such as RUP (Rational Unified Process).

The research required more in-depth analysis of interrelations and dependencies between Requirement Diagrams and other related SysML/UML diagrams, permitting system requirements to be translated into a precise system model. The results achieved enabled the establishment of a standpoint for conducting this investigation.

References

1. Ambler, S.W.: The Elements of UML 2.0 Style. University Press, Cambridge (2005)
2. Bajaj, A., Wrycza, S. (eds.): Systems Analysis and Design for Advanced Modeling Methods. Best Practices, Information Science Reference. IGI Global, New York (2009)
3. Bock, C.: SysML and UML 2 Support for Activity Modeling. Systems Engineering 9 (2006)
4. Booch, G., Rumbaugh, J., Jacobson, I.: The UML Reference Manual, 2nd edn. Addison-Wesley, Boston (2004)
5. Dennis, A., Wixom, B.H., Tegarden, D.: Systems Analysis and Design with UML Version 2.0 - An Object Oriented Approach, 2nd edn. Wiley, New York (2005)
6. Eriksson, H.E., Penker, M., Lyons, B., Fado, D.: UML 2 Toolkit. OMG Press, Indianapolis (2004)
7. Friedenthal, S., Moore, A., Steiner, R.: A Practical Guide to SysML. OMG Press, Indianapolis (2008)
8. Grady, R., Caswell, D.: Software Metrics: Establishing a Company-wide Program. Prentice Hall, New York (1987)
9. Hause, M.: SysML Hits the Home Straight (2009), http://www.esemagazine.com/index.php?option=com_content&task=view&id=140&Itemid=2
10. Larman, C.: Applying UML and Patterns: An Introduction to Object-Oriented Analysis and Design and Iterative Development, 3rd edn. Prentice Hall, New Jersey (2004)
11. Leffingwell D., Widrig D.: Managing Software Requirements. A Unified Approach. Addisson-Wesley, Boston (2000)
12. Maciaszek, L.: Requirements Analysis & System Design, 3rd edn. Addison-Wesley, Boston (2007)
13. Marcinkowski, B.: Business modeling with UML and BPMN: Features and Comparison. In: Proceedings of BIR 2008. The Seventh International Conference on Perspectives in Business Informatics Research. Gdansk University Press, Gdansk (2008)
14. Object Management Group: OMG Systems Modeling Language (OMG SysML). Version 1.2 (2010), http://www.omg.org/spec/SysML/1.2
15. Object Management Group: OMG Unified Modeling Language (OMG UML), Superstructure. Version 2.4 (2011), http://www.omg.org/spec/UML/2.4
16. Rumbaugh, J., Blaha, M.R., Lorensen, W., Eddy, F., Premerlani, W.: Object-Oriented Modeling and Design. Prentice Hall, New Jersey (1991)
17. Technical Board International Council on Systems Engineering: Systems Engineering Handbook. Version 3.2 (2010)
18. Weilkiens, T.: Systems Engineering with SysML/UML. Modeling, Analysis, Design. OMG Press, Indianapolis (2007)
19. Wojciechowski, A.: Introduction to requirements engineering (2009) (in Polish), www.inmost.org.pl/articles/Wprowadzenie_do_inżynierii_wymagań
20. Wrycza, S. (ed.): Proceedings of The Second AIS SIGSAND European Symposium on Systems Analysis and Design. Gdansk University Press, Gdansk (2007)
21. Wrycza, S., Marcinkowski, B.: A Light Version of UML 2: Survey And Outcomes. In: Proceedings of the 2007 Computer Science and IT Education Conference. University of Technology Mauritius Press, Mauritius (2007)
22. Wrycza, S., Marcinkowski, B.: The Language of Systems Engineering – SysML. In: Architecture and Applications. Helion, Gliwice (2009) (in Polish)
23. Wrycza, S., Marcinkowski, B., Wyrzykowski, K.: UML 2.0 in Information Systems Modeling. Helion, Gliwice(2005) (in Polish)

Part II

Integrated Systems Development

Customer Knowledge Management Models: Assessment and Proposal

Dorota Buchnowska

Department of Business Informatics
University of Gdansk
Sopot, Poland
dorota.buchnowska@univ.gda.pl

Abstract. In the current economy, customer knowledge is a key asset and a sustainable source of competitive advantage, which exerts an impact on the implementation of most processes in any organization. Efficient utilization of customer knowledge determines the development of a company. The main purpose of this paper is to present a review and assessment of presented in literature customer knowledge management models and to purpose an integrated management model for customer knowledge. This model should help in understanding how to manage customer knowledge in order to improve customer value.

Keywords: customer knowledge, customer knowledge management, CKM model, knowledge about customer, knowledge from customers, knowledge for customers.

1 Introduction

Knowledge has become a strategic resource as the basis of competitive advantage in an organization [2]. The most important type of knowledge would appear to be customer knowledge. The growing importance of customer knowledge is emphasized by numerous publications (e.g. [2], [3]) and is also confirmed by the empirical research. According to a survey conducted by Ernst and Young, customer knowledge was quoted as the most important type of knowledge (97%) to assist organizations act effectively. This is followed by knowledge about the best practices and effective processes (87%), and knowledge about competencies and capabilities (86%) [4]. Companies need a wide variety of knowledge about customers, such as: [5], [6]:

- who are their customers?
- how can they use knowledge to retain and support customers?
- how can knowledge help companies acquire new customers?
- how can companies use customer knowledge to continuously improve products and services?
- how can companies use customer knowledge to create new products and services?
- how can companies use customer knowledge to understand markets better?

To reach these levels of knowledge, most organizations have focused on collecting massive amounts of data about customers. Typically, they have a Customer Relationship

S. Wrycza (Ed.): SIGSAND/PLAIS 2011, LNBIP 93, pp. 25–38, 2011.
© Springer-Verlag Berlin Heidelberg 2011

Management (CRM) system that captures information about customer transactions [7]. However, such activities usually prove insufficient. A company collects large amounts of data and information but cannot transform them into customer knowledge, which would make the right business decisions.

Therefore, this research will present a model of customer knowledge management (CKM) which will help in understanding and organizing the processes associated with the creation and exploitation of customer knowledge while consciously managing them. Consequently, customer knowledge management becomes more efficient. To discuss the customer knowledge management model, it is necessary to define fundamental terms, such as customer knowledge (CK) and customer knowledge management (CKM).

2   The Concept of Customer Knowledge Management

2.1   The Concept of CK

Customer knowledge (CK) is a part of organizational knowledge. Zanjani, Rouzbehani and Dabbaghl [8] define it as "a kind of knowledge in the area of customer relationship, which has direct or indirect effect on our organizational performance". By the concept of knowledge, they also understand data or information which can be analyzed, interpreted and eventually converted to knowledge. However, information systems literature differentiates between the terms customer data, customer information, and customer knowledge [3], [9].

Data is collected in the organization's databases, paper documents and the minds of the employees. Typical customer data includes contact data, interaction data, purchasing data and customer feedback [10]. Nowadays, when customers willingly use their loyalty cards, dial into Voice Response Units and order off the Web, companies are awash in customer data. Most organizations have IT tools (such as CRM, Contact Center, ERP, e-business system) to gather customer data at every possible customer contact point [9]. These touch points include customer purchases, sales force contacts, service and support calls, Web sites visits, satisfaction surveys, credit and payment interactions or market research studies [11]. Every day, more and more transaction data is created. But collecting terabytes of customer transaction data does not guarantee business value. It is necessary to transform raw data into information and to integrate this information throughout the firm to develop knowledge useful in decision-making [12].

Information is data plus conceptual commitments and interpretations. Customer information is obtained through filtering, integrating, extracting or formatting customer data. As with gathering data, to transform customer data into customer information, organizations use various information systems. The most important of them are CRM, Business Intelligence and Customer Intelligence systems.

The concept of knowledge is actively and continually evolving. According to Nonaka [13], "information is a flow of messages, while knowledge is created and organized by the flow of information anchored on the commitment and beliefs of its holder". In this context, customer knowledge is information organized and analyzed so that it becomes understandable and applicable in solving problems and making decisions in the area of relations between an organization and its customers. Information can be transformed into knowledge through four processes: comparison, consequences, connections and conversation [14]. Types of customer information which serve as a source of customer knowledge are presented in [15]. Information technology can facilitate the accumulation of customer data and its transformation into customer information, but it is not able to convert information into knowledge, since knowledge is always related to a person or group of people [9], [16].

Despite the differences between the concepts data, information and knowledge, in this paper the term customer knowledge means knowledge and also data and information which can be converted into knowledge, because customer knowledge management includes management of data and information.

2.2 Kinds of Customer Knowledge

The relevant literature usually distinguishes between two kinds of customer knowledge [3], [17], [18]:

- knowledge about customers that relates to the customer, but is owned by the organization; this may include knowledge about actual and potential customers and customer segments, as well as knowledge about individual customers;
- knowledge possessed by customers and potential customers, including knowledge about product ranges, companies, and the marketplace.

However, according to Gebert et al. [19] customer knowledge can be classified into three categories:

- knowledge about customers,
- knowledge from customers,
- knowledge for customers.

Knowledge about customers is accumulated in order to know customers better, to understand their expectation, needs and motivations and to address them in a personalized way [17]. This includes customer histories, connections, requirements, purchasing activity, buying habits and payment behavior [19]. This kind of customer knowledge is acquired mainly in a passive way, i.e. not actively by interaction with the customer. It is the result of analyses, interviews and observations as conducted, for example, by market research institutions [20]. Companies not only gain knowledge about customers but also purchase data, information and knowledge about customers [8].

Knowledge from the customer is the knowledge that organizations receive from its customers. This category of knowledge includes: customer's knowledge of products, suppliers and markets [19], their ideas and recommendations concerning the improvement of the product [21], ideas, thoughts and information regarding the preferences, creativity or experience with products, services, processes or expectations [17]. Knowledge from customers is valuable as it leads to measures which improve products and services. Efforts need to be made to channel this knowledge back into the enterprises. [1], [22] This type of customer knowledge mostly arrives at the company in a direct way [20].

Knowledge for the customer is required to satisfy the knowledge needs of customers [22], [2]. Examples include knowledge about products, markets and suppliers [23] This kind of knowledge can be gained from other customers, information consulting institutes, competitors and the company itself [1], [22]. Customers should be supported with "knowledge for the customer" during the entire buying cycle. Daneshgar and Bosanquet [4] assert that knowledge for the customer is a product of the integration and transformation of the previous two categories of customer knowledge. Table 1 shows the different types of customer-oriented knowledge and their typical contents.

Table 1. Comparison of types of customer knowledge [20]

	Knowledge about Customer	Knowledge from Customer	Knowledge for Customer
Company/ person	• B2B: industry credit worthiness • B2C: age, sex, income…	• Own objectives, strategies, own expectations, interests…	• Specify problems and ascertain demand
Product/ service	• Product portfolio, purchase history, contract durations…	• Strengths/weaknesses of quality compared to competitor	• Scope of offer, quality features, prices…
Actions of company	• Type, intensity, frequency of customized activities	• Strengths/weaknesses of activities compared to competitor	• Special offers, individual talks, special conditions…
Reactions of customer	• Turnover, gross margin, customer lifetime value, customer satisfaction, complaints	• Insights and intentions concerning products and services	• Archived customer status (e.g. in customer binding programs) or discount stages

Smith and McKeen [5] suggest that there is one other type of customer knowledge which is co-created knowledge, which comes about as the result of cooperation between the organization and its clients and generates value for the customer as well as for the organization. Differences between different kinds of knowledge flows are illustrated in Fig. 1.

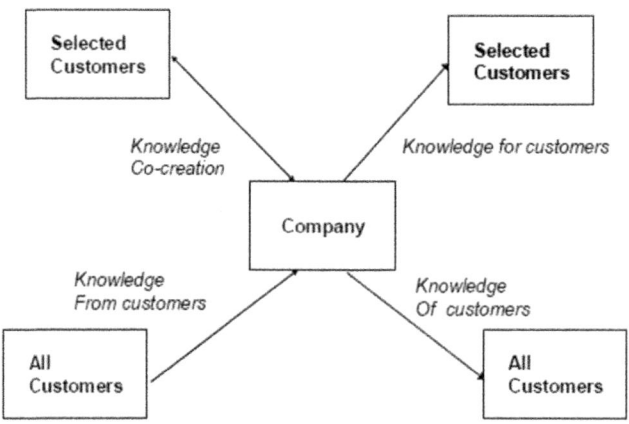

Fig. 1. Customer knowledge flows [5]

The evolution of customers from a passive audience for a firm's offers to active players means that the value derived from knowledge co-creation with customers around developing innovative products is becoming increasingly important [24]. Table 2 shows a summary of the literature review of the types of customer knowledge.

Table 2. Taxonomy of customer knowledge

Author	Types of CK
Rowley (2002)	Knowledge about customers
	Knowledge possessed by customers
Gebert et al (2002)	Knowledge for customer
	Knowledge from customer
	Knowledge about customer
Smith, McKeen (2005)	Knowledge for customer
	Knowledge from Customer
	Knowledge of customer
	Co-created knowledge

In addition to the taxonomy presented, customer knowledge can be further subdivided by division criteria used for the classification of knowledge in general. Customer knowledge can be explicit (the structured information in databases) or tacit customer (knowledge in the minds of employees and customers) [25] and individual or collective. Finally, customer knowledge can be stratified in different levels ranging from lowest to highest: cognitive (know-what), conditional (know-when), relational (know-with), applied (know-how) and rule-of-thumb (know-why) [6]. The current state of technology for supporting knowledge management systems suggests that completely different types of component technologies and methods are required for the management of specific types of knowledge, in particular tacit and explicit knowledge [4].

2.3   The Concept of CKM

There is no single, universally accepted definition of customer knowledge management (CKM). This follows from the fact that we can meet with different interpretations of the concept of customer knowledge in the literature. CKM is usually defined as an ongoing process of generating, disseminating and using customer knowledge within an organization, and between an organization and its customers [9]. However, CKM may also concern the management and exploitation of different types of customer knowledge:

- with knowledge residing on the customer side [26], [27]
- with knowledge about customers [18],
- with two types of customer knowledge: knowledge possessed by customers and knowledge about customers [17],
- and with the knowledge about customers, knowledge for customers and knowledge from customers [12], [8].

Dobney [28] rightly stresses that customer knowledge can be approached from two ends. Firstly, it can be said that customer knowledge refers to the knowledge that an organization possesses about its customers. An alternative interpretation of customer knowledge is that it is the knowledge that a company needs to have to build stronger customer relationships. Using the first definition, the role of customer knowledge management is to capture and organize data, information and knowledge about customers in order to allow it to be shared and discussed throughout the organization. From the second point of view, what a company currently knows about customers is not sufficient. CKM, in this case, primarily refers to processes and systems to gather more information and data about who customers are, what they do and how they think.

Customer knowledge management differs from traditional knowledge management (KM). "Whereas traditional knowledge management is about efficiency gains (avoiding of re-inventing the wheel), CKM is about innovation and growth" [8]. The differences between KM and CKM are presented in Table 3.

Table 3. Knowledge Management versus CKM [27]

	KM	CKM
Knowledge sought in	Employee, team, company, network of companies.	Customer experience, creativity, and (dis)satisfaction with products/services.
Axiom	"If only we knew what we know".	"If only we knew what our customers know."
Rationale	Unlock and integrate employees' knowledge about customers, sales processes and R&D.	Gaining knowledge directly from the customer, as well as sharing and expanding this knowledge.
Objectives	Efficiency gains, cost saving and avoidance of reinventing the wheel.	Collaboration with customers for joint value creation.
Metrics	Performance against budget.	Performance against competitors in innovation and growth, contribution to customer success.
Benefits	Customer satisfaction.	Customer success, innovation, organizational learning.
Recipient of Incentives	Employee.	Customer.
Role of customer	Passive, recipient of product.	Active, partner in value-creation process.
Corporate role	Encourage employees to share their knowledge with their colleagues.	Emancipate customers from passive recipients of products to active co-creators of value.

In literature, CKM is often erroneously identified as Customer Relationship Management (CRM), which has been defined as "the business strategy, process, culture and technology that enables a company to optimize revenue and increase value through understanding and satisfying the individual customer's needs" [29]. CRM integrates people, process and technology to maximize relationships with all customers [30]. The primary difference between CRM and CKM is that CRM is largely focused on knowledge about the customers, while CKM first and foremost focuses on knowledge from the customers [20], [27]. Liyun et al. [31] note many more differences which are shown in Table 4.

Table 4. Customer Knowledge Management versus Customer Relationship Management [31]

	CRM	CKM
Communication manner	Unilateralism, structure data and information	Bidirectional, non-structure
Knowledge sought	Customer database	customer database, customer experience, creativity
Sort of customer knowledge	Knowledge about customers	Knowledge reside in customers
Rational	Mining knowledge about the customer in company's database	Gaining knowledge directly from the customer, as well as sharing and expanding this knowledge
Recipient of incentives	Customer	Customer and employee
Benefits	Customer retention	Collaboration with customers for joint value creation
Valuation criteria	Customer satisfaction and customer loyalty	Ability of creation and customer satisfaction
Role of customer	Captive, tied to produce/service, by loyalty schemes	Active, partner in value-creation process
Corporate role	Building lasting relationships with customers	Emancipate customers from passive recipients of products to active co-creator of value

A number of authors (e.g. [19], [9]) see CKM as the result of the integration of KM and CRM, while others consider that the result of such integration is Knowledge-enabled Customer Relationship Management (KCRM) [32], [33].

In this paper, CKM means a comprehensive approach to customer knowledge ([2], [34] present a similar approach). It includes more than just knowledge from customers but also knowledge about and for customers. CKM uses KM methods and tools to effectively manage customer-oriented knowledge. An effective CKM system requires (by analogy with knowledge management) the integration of five essential elements: business strategy, culture, people, processes and technology.

3   A Review and Assessment of CKM Models

3.1   Conceptual CKM Model

In recent years, a growing interest in customer knowledge management has been observed. The appearance of several theoretical CKM models results from research in this field. Two of these deserve attention: the conceptual model proposed by Zanjani, Rouzbehani and Dabbagh [8] and the process model proposed by Gebert et al. [19].

Zanjani, Rouzbehani and Dabbagh [8] suggest a customer knowledge management conceptual model to help companies understand distinctive types of customer knowledge. This is shown in Figure 2.

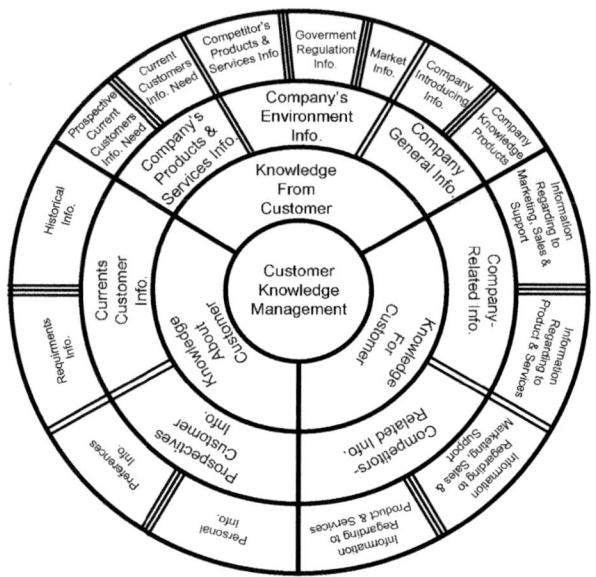

Fig. 2. Conceptual model of Customer Knowledge Management [8]

They distinguish three types of customer knowledge, similar to other authors' statements in CKM literature: knowledge for customer, knowledge from customer and knowledge about customer (the first layer of the model) and they break each type of customer knowledge down into more detailed knowledge-based parts (the second and third layers of the model presented). Presented model systematizes the knowledge of customer types but does not show the relationships between different types of knowledge. This, and almost all KM models at the conceptual level, can be traced back to a basic approach to analyze what knowledge is and how it is created. Consequently, from a business practice point of view, a common weakness of CKM models is that knowledge is separated from relevant business processes [9].

3.2   Process Oriented CKM Model

To fill this gap Gebert et al. [19] present a process oriented CKM model (Fig. 3). This model is based on the integration of CRM and KM approaches. The authors claim that by integrating both approaches into a customer knowledge management model, the benefit of using CRM and KM can be enhanced and the risk of failure in this kind of project reduced [23].

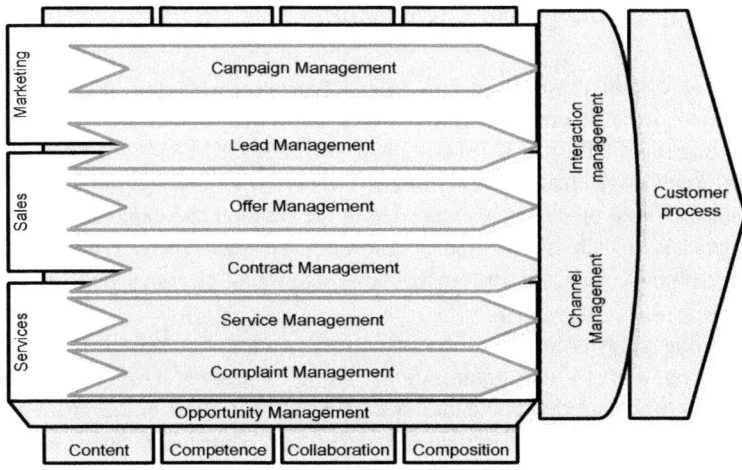

Fig. 3. A customer oriented knowledge management model [19]

The CKM process model as introduced by Gebert et al. [23] offers a process perspective to illustrate which KM tools can be applied to the CRM sub-processes to achieve effective CKM [17]. The model identifies six business processes: campaign management, lead management, offer management, contract management, complaint management and service management [6]. It introduces the four KM aspects of content, competence, collaboration and composition. These aspects were arrived at by analyzing existing KM models and numerous case studies [2]. The first aspect (content) focuses on the management of explicit knowledge, and is supported by technologies such as content management or document management system. Competence refers to the management of both implicit and explicit knowledge [35] and focuses on understanding which customer knowledge is required by employees in order to accomplish their tasks in different business processes with customers [9]. Competence is supported by technologies such as expertise directories, skill management and e-learning systems [35]. Collaboration deals with the creation and dissemination of knowledge among a few individuals, e.g. in project teams. Typical functions that support the aspect of collaboration are: email, group information tools, and instant messaging systems [2]. The final aspect, composition, refers to the management of dissemination and usage of knowledge among a large number of individuals in an organization [35]. This requires technologies such as search and navigation and uses systems such as knowledge mining systems, personalization, taxonomy management systems and knowledge maps [2].

The model presented by Gebert focuses on the processes associated with managing customer relationships. Knowledge management should primarily provide support to the implementation of these processes. The same applies to other models of CKM presented in the literature, for example in [38].

4 Proposal of Integrated CKM Model

Despite their cognitive values, the presented models do not pretend to explain how customer knowledge is created, transferred or developed. Therefore, based on existing literature from the fields of KM (in particular [36]) and CKM [37] this paper proposes a customer knowledge management model which will help in understanding the relationship between processes associated with the creation and exploitation of customer knowledge and which shows the relationship between three types of customer knowledge: knowledge from customer, knowledge about customer and knowledge for customer (Fig. 4).

The starting point is to define the objectives of customer knowledge management which are relevant to the organization. These objectives are the foundation of identification of knowledge about the customer. Identification is the process whereby sources of knowledge are identified. This process consists of three sub-processes: defining the scope of knowledge required about customers, indicating gaps in knowledge about customers and identifying sources and methods of knowledge acquisition. This is designed to provide information relevant to the implementation of predefined goals and to simultaneously avoid information overload.

The next process, acquisition of customer knowledge, is a process of knowledge transfer from the environment of the organization to its interior, and the process of acquiring knowledge from internal sources. Nowadays, particular importance is placed on the process of acquiring knowledge from customers. Smart companies realize that customers are more knowledgeable than they might think and consequently seek knowledge through direct interaction with them, in addition to seeking knowledge about them from their sales representatives [27] or from databases. The knowledge acquired should be codified and stored in corporate databases and knowledge warehouses.

Development of customer knowledge is complementary to knowledge acquisition. It focuses on generating new skills, new products, better ideas to meet customer needs more effectively and improving processes in order to increase the value offered to the customer. Distribution is the process of sharing and spreading customer knowledge already present within the organization. Processes of customer knowledge development and distribution require an organizational culture that supports organizational learning, collaboration with clients and sharing of knowledge with other workers and customers.

Customer knowledge utilization is the next process in the knowledge cycle. Previously gained knowledge (from and about customers) must be applied in order for the organization to successfully meet organizational goals and objectives. In this phase knowledge for the customer is produced, a knowledge essential in developing relationships with customers, increasing the level of customer satisfaction and loyalty and consequently improving the competitiveness of the enterprise.

We must remember that customer knowledge management is not a single action. All the processes must be continuously pursued and the practicality of customer knowledge should be verified. Effectiveness of customer knowledge management is significantly increased when each process of CKM is supported by appropriate information technologies.

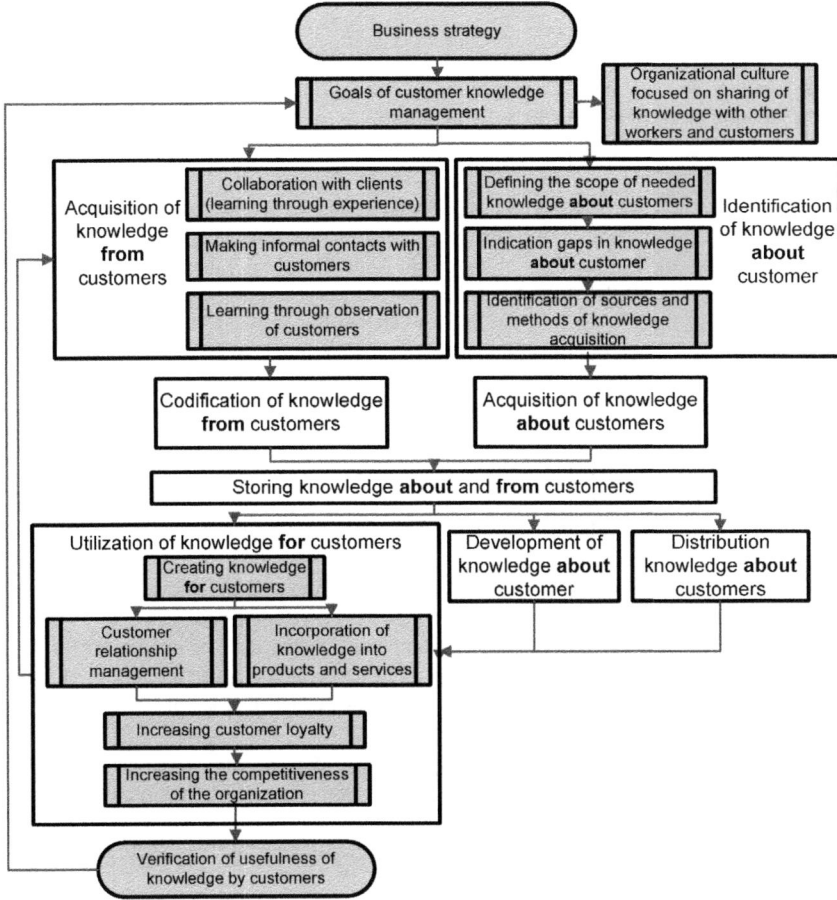

Fig. 4. Customer knowledge management model

5   Conclusion

This paper presents an integrated model of customer knowledge management which helps businesses to understand how to manage of all kinds of customer knowledge so as to improve its competitiveness.

Work in this direction will be continued. The model will be supplemented by an analysis of the usefulness of various information technologies in the individual processes of customer knowledge management. The proposed model is based on existing literature from the fields of knowledge management and customer knowledge management . The theoretical model will be verified by empirical research.

References

1. ALHawari, S., Talet, A.N., Mansou, E., Alryalat, H., Hadi, W.M.: The Impact of Knowledge Process about Customer on the Success of Customer Knowledge Acquisition. Communications of the IBIMA 1, 27–33 (2008),
http://www.ibimapublishing.com/journals/CIBIMA/volume1/v1n3.pdf
2. Bueren, A., Schierholz, R., Kolbe, L., Brenner, W.: Customer knowledge management - improving performance of customer relationship management with knowledge management. In: Proceedings of the 37th Annual Hawaii International Conference on System Sciences. IEEE (2004)
3. Rowley, J.E.: Reflections on customer knowledge management in e-business. Qualitative Market Research: An International Journal 5(4), 268–280 (2002)
4. Daneshgar, F., Bosanquet, L.: Organizing Customer Knowledge in Academic Libraries. Electronic Journal of Knowledge Management 8(1), 21–32 (2010)
5. Smith, H.A., McKeen, J.D.: Developments in Practice XVIII – Customer Knowledge Management: Adding Value for Our Customers. Communications of the Association for Information Systems 16, 744–755 (2005)
6. Al-Shammari, M.: Customer Knowledge Management: People, Processes, and Technology. IGI Global, London (2009)
7. Ogunde, A.O., Folorunso, O., Adewale, O.S., Ogunleye, G.O., Ajayi, A.O.: Towards an Agent-Based Customer Knowledge Management System (ABCKMS) in E-Commerce Organizations. International Journal on Computer Science and Engineering 02(06) (2010)
8. Zanjani, M.S., Rouzbehani, R., Dabbagh, H.: Proposing a Conceptual Model of Customer Knowledge Management: A Study of CKM Tools in British Dotcoms. World Academy of Science, Engineering and Technology (38) (2008)
9. Rollins, M., Halinen, A.: Customer Knowledge Management Competence: Towards a Theoretical Framework. In: Proceedings of the 38th Hawaii International Conference on System Sciences (2005),
http://www.computer.org/comp/proceedings/hicss/2005/2268/08/22680240a.pdf
10. Wrycza, S. (ed.): Informatyka dla ekonomistów. Podręcznik akademicki. PWE, Warszawa (2010)
11. Kotler, P., Armstrong, G.: Principles of marketing. Pearson Prentice Hall, New Jersey (2010)
12. Belbaly, N., Benbya, H., Meissonier, R.: An empirical investigation of the customer Knowledge creation impact on NPD Performance. In: Proceedings of the 40th Hawaii International Conference on System Sciences (2007)
13. Nonaka, I.: A dynamic theory of organizational knowledge creation. Organization Science 5(1), 14–37 (1994)
14. Davenport, T., Prusak, L.: Working Knowledge. Harvard Business School Press, Boston (1998)
15. Buchnowska, D.: Lokalizacja wiedzy o kliencie. In: Oniszczuk-Jastrząbek, A., Gutowski, T., Żurek, J. (eds.) Przedsiębiorstwo na Rynku Globalnym. Fundacja Rozwoju Uniwersytetu Gdańskiego, Gdańsk (2010)
16. Ziemba, E., Minich, M.: Informacja i wiedza w przedsiębiorstwie. In: Oleński, J., Olejniczak, Z., Nowak, J. (eds.) Informatyka. Strategie i zarządzanie wiedzą, Polskie Towarzystwo Informatyczne, Katowice (2005),
http://www.ae.katowice.pl/images/user/File/katedra%20informatyki%20ekonomicznej/PTI2005.pdf (July 02, 2011)

17. Peng, J., Lawrence, A., Lihua, R.: Customer Knowledge Management in International Project: A Case Study (2011),
http://motsc.org/Customer_Knowledge_Management_in_Internatio
nal_Projectwithquestions.pdf
18. Xu, M., Walton, J.: Gaining customer knowledge through analytical CRM. Industrial Management & Data Systems 105(7), 955–971 (2005)
19. Gebert, H., Geib, M., Kolbe, L., Riempp, G.: Towards Customer Knowledge Management: Integrating Customer Relationship Management and Knowledge Management Concepts. In: Proceedings of ICEB Conference, Taiwan (2002)
20. Wilde, S.: Customer Knowledge Management. Improving Customer Relationship Through Knowledge Application. Springer, Heidelberg (2011)
21. Triki, A., Zouaoui, F.: Customer Knowledge Management Competencies Role in the CRM Implementation Project. Journal of Organizational Knowledge Management 2011 (2011),
http://www.ibimapublishing.com/journals/JOKM/2011/235827/235
827.pdf
22. ALHawari, S., Talet, A.N., Mansou, E., Alryalat, H., Hadi, W.M.: The effects of Knowledge Process for Customer on the achievement of Customer Knowledge Retention. Communications of the IBIMA 6, 174–180 (2008),
http://www.ibimapublishing.com/journals/CIBIMA/volume6/
v6n27.pdf
23. Gebert, H., Geib, M., Kolbe, L., Brenner, W.: Knowledge-enabled customer relationship management integrating customer relationship management and knowledge management concepts. Journal of Organizational Knowledge Management 7(5), 107–123 (2003),
http://j.pelet.free.fr/publications/km/
Knowledge-enabled_customer_relationship_management_
integrating_customer_relationship_management_and_knowledge
_management_concepts.pdf
24. Lawer, C.: On customer knowledge co-creation and dynamic capabilities (2005),
http://www.customerthink.com/blog/on_dynamic_capabilities
_and_customer_knowledge_co_creation
25. Polanyi, M.: The Tacit Dimension. Doubleday, Garden City (1966)
26. Zhang, Z.: Customer knowledge management and the strategies of social software. Business Process Management Journal 17(1), 82–106 (2011)
27. Gibbert, M., Leibold, M., Probst, G.: Five Styles of Customer Knowledge Management and How Smart Companies Put them into Actions. European Management Journal 20(5), 459–460 (2002)
28. Dobney: What is customer knowledge (2011),
http://www.dobney.com/Knowledge/ck_definition.htm
29. Reynolds, J.: A practical guide to CRM: building more profitable customer relationships. CMP Books, New York (2002)
30. Goldenberg, B.J.: CRM Automation. Prentice Hall PTR (2003)
31. Liyun, Q., Keyi, W., Xiaoshu, W., Fangfang, Z.: Research on the Relationship among Market Orientation, Customer Relationship Management. Customer Knowledge Management and Business Performance, Management Science and Engineering 2(1), 31–37 (2008)
32. Tiwana, A.: The Essential Guide to Knowledge Management: e-Business i CRM Applications. Prentice Hall PTR (2001)
33. Al-Shammari, M.: Implementing Knowledge-Enabled CRM Strategy in a Large Company: A Case Study from a Developing Country. In: Jennex, M.E. (ed.) Case Studies in Knowledge Management. Idea Group Publishing (2005)

34. Desouza, K.C., Awazu, Y.: Gaining a competitive edge from your customers. Exploring three dimensions of customer knowledge. KM Review 7(3), 12–15 (2007)
35. Chan, J.O.: Integrating knowledge management and relationship management in an enterprise environment. Communications of the IIMA (2009),
    http://findarticles.com/p/articles/mi_7099/is_4_9/
    ai_n56337597/pg_4/
36. Probst, G., Raub, S., Romhardt, K.: Zarządzanie wiedzą w organizacji. Oficyna Ekonomiczna, Kraków (2002)
37. Buchnowska, D.: Propozycja modelu zarządzania wiedzą o klientach. In: Research Papers of Economic University of Szczecin (in print, 2011)
38. Dous, M., Salomann, H., Kolbe, L., Brenner, W.: Knowledge Management Capabilities in CRM: Making Knowledge For, From and About Customers Work. In: Proceedings of the Eleventh Americas Conference on Information Systems, Omaha, NE, USA, pp. 167–178 (2005)

ERP in the Cloud – Benefits and Challenges

Anna Lenart

University of Gdansk, Department of Business Informatics, Sopot, Poland
anna.lenart@ug.edu.pl

Abstract. Cloud computing (CC) is one of the most important revolutionary changes in Information and Communication Technology (ICT). Its roots can be traced to technology and business trends. Nowadays organizations have three ERP system deployment scenarios: on-premise, hosting and on-demand (EaaS). The aim of the paper is to analyse the benefits and the challenges of Cloud ERP in the SaaS model. Firstly, SaaS as a business model of cloud computing is described. The issues concerning traditional and Cloud ERP are also discussed.

Keywords: cloud computing (CC), Software as a Service (SaaS), Enterprise Resource Planning (ERP), ERP as a Service (EaaS), SaaS ERP, Cloud ERP.

1 SaaS as a Business Model of Cloud Computing

1.1 Foundation of Cloud Computing

The term 'cloud computing' has a different meaning for IT developers, IT managers or end users. It can be described as "Web-based applications that are stored on remote servers and accessed via the 'cloud' of the Internet using a standard Web browser" [1]. For Gartner CC "is a style of computing where massively scalable IT-related capacities are provided 'as a service' across the Internet to multiple external customers" [2]. The cloud service provider (CSP) evolved from the internet service provider (ISP) [3]. CC is based on services (for example access to ERP system) and requires almost no IT investment. It also can be a way of IT cost optimization and fast delivery of new IT systems.

The United States National Institute of Standards and Technology (NIST) has defined five characteristics of cloud computing [4]: on-demand self-service, broad network access, location independent resource pooling and payment for resources used, rapid elasticity and measured service. This concept promises to cut capital and operational costs and let IT departments focus on strategic projects instead of on operational activities.

Clouds will become more and more important in the ICT world and may change it in a similar way to the Internet. The roots of CC can be found in technology and business trends [5]. The foundational enabler of cloud computing is the convergence of various IT technologies [4]: hardware virtualization, distributed computing (grid computing, utility computing), Internet technology (SOA, Web services, Web 2.0, broad-band networks), system management (service level agreements, data centre automation) and open source software.

S. Wrycza (Ed.): SIGSAND/PLAIS 2011, LNBIP 93, pp. 39–50, 2011.

Technology trends refer to the virtualization of IT resources used in grid computing and utility computing concepts. In CC, the computer is a set of many virtual machines. Cloud is the concept of offering the customer a virtual cloud of service. This customer does not have to know the details of the service, as they are not required when using it. In cloud computing, as opposed to grid computing, resources follow customer needs. A further concept is utility computing, known as on-demand computing. By using this concept, an organization can reduce their investment in IT infrastructure. The technology trend is also related to service-oriented architecture (SOA).

Business trends can be associated with the acquisition of IT resources when necessary and the decomposition of value chain and transformation of business networks (for example, a virtual organization). The most important reason for the rise in CC's popularity is the search for an adaptive and dynamic IT infrastructure which does not hinder business development. Another business trend is the concept of business process management (BPM), which facilitates integration and communication of heterogeneous IT environments. When an organization is planning to rent IT resources and pay for IT services, it has to know which business process needs IT support and how this can be achieved. The provider should ensure the right level of IT services according to the service level agreement (SLA).

1.2 Ontology of Cloud Computing

Clouds have a specific, five-layered structure [4], [6-7]. In order to create an application, it is necessary to have an infrastructure composed of three layers: the hardware, the operating system and a cloud software infrastructure. The hardware - HaaS (Hardware as a Service) is the lowest layer, which provides essential power capacity to the customer or the software developer. The operating system should be suitable for the creation of a cloud environment. A cloud software infrastructure is a set of tools necessary for the management of the power capacity - IaaS (Infrastructure as a Service) and the disk space - DaaS (Data Storage as a Service), as well as communication - CaaS (Communication as a Service).

Software developers use the fourth layer as the cloud software environment for building the software in the cloud. It can be PaaS (Platform as a Service). The customers use the fifth layer which is cloud application. It can be SaaS (Software as a Service). The SaaS model means "software that is delivered on-demand over a network (…) through a rental ownership model" [8]. The pioneer of SaaS is Salesforce.com (CRM system).

SaaS, PaaS and IaaS are the three main models of CC delivery (service model) and parts of the CC market. They are known as the SPI model, which is the basis of cloud architecture implementation. It contains elements of IT systems and their connections with XaaS (X as a service) services [9]. In the IaaS model, customers rent only the computer infrastructure. In the PaaS model, both infrastructure and programming tools are hosted by the vendor. The platform can be used for the implementation of a Web-based application on the hosted infrastructure. In the SaaS model, customers pay for software hosted by the vendor. SaaS is not the same as the Application Service Provider (ASP) known in IT outsourcing. Services are provided to multiple customers in both models, but the SaaS provider does not offer dedicated infrastructure [3] and the customer does not know where the software is located. SaaS is a kind of IT e-outsourcing solution which can reduce the cost of service delivery.

1.3 Deployment Models of Cloud Computing

It is also necessary to describe CC deployment models. A brief description of four deployment models of cloud computing is presented in Table 1.

Table 1. Cloud computing deployment models (Source: Based on [2], [4], [6-7])

Model	Characteristics
Public cloud	— the infrastructure is the property of one organization which sells cloud services for the society or a specific sector of the economy; — the services are accessible by Internet; — this model is suitable for small and mid-size enterprises; — can be used for business applications like SaaS ERP in order to reduce capital expense
Private cloud	— the infrastructure is the property of or rented by the organization and is used only by this organization (internal cloud); — the services are accessible and managed in a corporate network and access to the cloud can be limited to one department or cost centre in order to provide control; — this model is preferred by multidivisional enterprises or international corporations and is also suitable for business applications like the traditional ERP system; — cannot be used to deliver applications in the SaaS model
Community cloud	— is used by many organizations and supports communities which have common goals; — can be used for communication between members of a project team (groupware applications)
Hybrid cloud	— is the combination of a minimum of two models of cloud (public, private or community) which form unique units although connected by one technology; — the enterprise can prefer a different model of CC in order to use various software categories

For a small or mid-size enterprise (SME), a public cloud is the chance to have more flexible and accessible IT systems. In times of prosperity or growth in the enterprise, it is possible to increase the power capacity of the IT environment. In times of crisis, the enterprise can decrease the range of resources or services used. Examples of public clouds are: Amazon Elastic Cloud Compute (E2C), Google Apps Engine and Microsoft Azure.

1.4 SaaS Market in the World

The SaaS is the most popular CC delivery model. The cloud technology and web-based software make SaaS effective to deliver, but an organization can have SaaS without cloud technology. Cloud-based technology can be used to improve hosting and deliver a SaaS environment.

According to Gartner Research in 2009 (Fig. 1), enterprises around the world have purchased the SaaS product to a level of almost 7.5 billion US dollars. These enterprises have most often bought tools for remote collaboration, file sharing and communications or web conferencing, for example, (2.57 billion US dollars). Next come CRM systems (2.28 billion US dollars) and ERP systems (1.23 billion US dollars). Other categories of business solutions (including SCM systems) have generated an income of 1.43 billion US dollars. Enterprises are not very interested in office software packages like Microsoft Office offered in the SaaS model (68 million US dollars) [7].

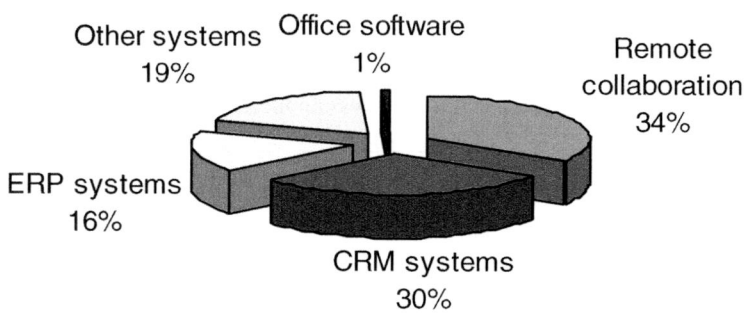

Fig. 1. The popularity of business solutions in the SaaS model in 2009 (Source: drawn up on the basis of [7])

By reference to IDC in 2010, the value of the SaaS service market has reached 9.2 billion US dollars and increased by 15.7 % in comparison to 2009. In 2011, the value of this market will reach 10.7 billion US dollars - an increase of 16.2%. The SaaS market in 2010 achieved 10% of the whole software market. In 2012, 85% of new software will be delivered in the cloud [10]. Gartner forecasts that in spite of limited trust of the SaaS model, the sale of software in this model will grow systematically and in 2014 will reach 14 billion US dollars globally [7]. In Europe, SaaS is not as popular as elsewhere in the world. In Poland about 100 vendors offer applications in the SaaS model, but there are no SaaS market statistics [10].

2 Issues Concerning Traditional and Cloud ERP

2.1 ERP as an Integrated Management Information System

The term Enterprise Resource Planning (ERP) system dates from 1990 when Gartner used it for the first time. ERP is a kind of information system which is considered as a cross-functional, process-oriented and legacy system because it integrates management information across the entire enterprise and also serves the information needs of the entire enterprise.

Nowadays, the ERP system is the "paradigm of organizational computing" [11] and is referred to as "the set of activities that managers used to run the important parts of an organization such as purchasing, human resources, accounting, productions and sales" [11]. The ERP system is the backbone of information systems in an enterprise or financial and government institution [12]. The traditional ERP system for large enterprise consisted of many core modules grouped by business areas such as: Finance Resource Management (FRM), Human Resource Management (HRM), Supply Chain Management (SCM), Customer Relationship Management (CRM). Additional modules which can be integrated within the ERP system include: Project Management System (PMS), Enterprise Performance Management (EPM) and Governance, Risk and Compliance (GRC).

The ERP system is the most risky, time-consuming and expensive IT investment in the enterprise. It is sold in modules or functional components and is employed by many users involved in business processes who share information across departments and offices or with business partners. The enterprise does not have to implement each module, but more modules lead to greater integration and a return on investment. The success of the ERP system implementation requires [11]: organizational changes, business process modification, user training and motivation in all employees.

The main ERP systems trends predicted for 2011 were [13]:

- more enterprises realize that there are viable alternatives to the traditional ERP system,
- clouds co-exist with traditional software deployments,
- ERP will get serious in the cloud,
- vendors will built new systems to take advantage of cloud computing.

ERP is a core business system and should be tailored to the enterprise's needs. A traditional client-server ERP system is usually customized software, while ERP as a Service (EaaS) model is package software. In the case of client-server software, the applications and storage are centralized and have to be upgraded on the client's desktop. EaaS eliminates many barriers to implementing or upgrading an ERP system. It is a way of accessing innovative technology that will lower current internal IT resources and maintenance costs.

2.2 ERP Systems Deployment Scenarios

Nowadays organizations have three ERP system deployment scenarios. An organization has the option of purchasing a license or an SaaS ERP solution. When a license is purchased, it can deploy the legacy ERP system in its own data centre (on-premise) or can outsource operations to an external provider (hosting off-site). An on-premise ERP system solution is known to be safer and more reliable. In the case of an SaaS ERP solution, the organization rents "a complete turnkey package that includes software and the entire delivery mechanism" [14]. Renting a SaaS ERP can turn out more expensive in the long term than purchasing the software license. The more important thing in the SaaS ERP model is reliance on an external provider. An on-premise solution can be used in a private cloud. The hosted ERP is good for a private or public cloud but SaaS ERP is only good for a public cloud. A modern ERP offers the switch between public and private cloud [15].

According to a Panorama survey [16], 17% of enterprises in 2010 implemented SaaS for ERP solutions, as opposed to a mere 6% in 2009. Other deployment models achieved a greater share of the ERP market: on-premise ERP (59%) and traditional

ERP hosted off-site (24%). ERP as a Service (EaaS) still has a small adoption rate and represents only a small percentage of the SaaS market. In mid-2010 the willingness to consider SaaS ERP among small and mid-size enterprises (SME) rose to 39% from 26% in 2007 [17].

An organization can find advantages and disadvantages in acquiring an ERP system on-premise or as a service. The decision depends on the needs, for example, how frequently business requirements are changing and if a flexible software model is required. Another problem is the possibility of shifting between SaaS ERP and the on-premise model [8]. The portability of applications and data outside the cloud is a big problem. Other important factors in choosing an ERP deployment scenario are company size, compliance with law and security risk.

2.3 Types of Cloud ERP and Their Uses

The Cloud ERP offers the customer speed of implementation and lower costs of entry. It is the shortest possible route to a new ERP system. There is a difference between Cloud ERP and SaaS ERP (EaaS): "Cloud ERP is hosted service delivered over the Internet" [18]. The ERP system in the EaaS model resides in the cloud, which provides computing power to run the ERP system. The system is available to the user on-demand once the subscription fee is paid. For secure access, the user needs an Internet connection. SaaS is not a required component of ERP software but organizations can purchase the more flexible Cloud ERP system when it is offered in a SaaS model. An organization can have Cloud ERP without SaaS (cloud infrastructure or cloud platform), SaaS ERP without cloud (web-based ERP) or SaaS ERP enabled by cloud (cloud application) [19]. Different types of Cloud ERP are illustrated on Fig. 2.

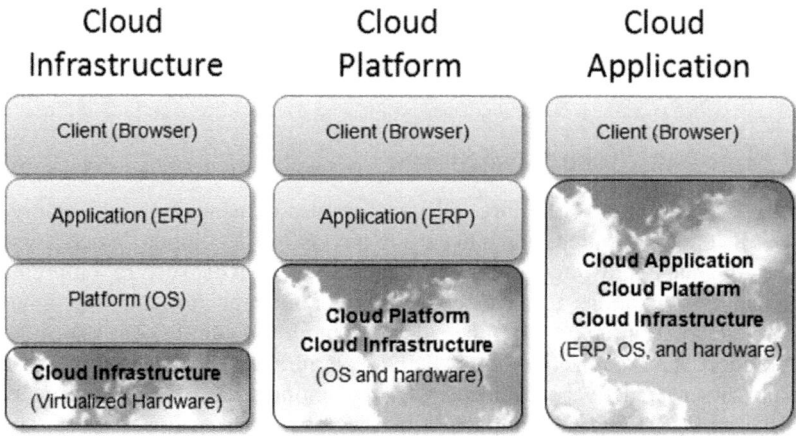

Fig. 2. Types of Cloud ERP (Source: based on [20])

It is also possible for legacy ERP software to be delivered on a hosted virtual private server as cloud software. An ERP system on the cloud, "hosted and accessed using a VPN or client software is benefiting from the cloud bandwagon but not achieving the true benefits of Cloud ERP" [21]. A better way is not to use a legacy

ERP in the new cloud infrastructure, but in order to gain maximum benefits of CC, to use a modern Cloud ERP application instead.

Cloud ERP is not the best choice for every enterprise. It is necessary to answer the question of what the Cloud ERP object is for. First of all, Cloud ERP can be a good solution for enterprise just starting its business activity and which does not want to pay for business software. Next, Cloud ERP is good solution for a multidivisional enterprise which has a distributed structure and wants to begin activity in a new division or shop without the investments in IT infrastructure. Another example is an enterprise which has an on-premise ERP system but which was acquired by another software vendor and the ERP system will no longer be supported. Cloud concept can also be a solution for using applications which the enterprise does not have in house, for example, additional modules of ERP systems like BI (Business Intelligence). Cloud ERP is a revolutionary change ideal for innovative organizations. It gives small and mid-size enterprise a chance, in that they can gain access to advanced business software which they are unable to purchase as an on-premise solution.

2.4 Vendors of Cloud ERP in the SaaS Model

There is some confusion on the CC market because many vendors do not distinguish between web- and cloud-based software [21]. Neither has to be upgraded on the client desktop because each is used by the web browser. Web-based software is easily accessed by employees, suppliers, clients and business partners and even by mobile devices. Cloud ERP can be achieved with or without web-based software but web-based is critical. Cloud solutions of Infor and Sage, for example, do not utilize web-based software. In order to deliver web-based cloud vendors, new product lines such as Epicor Express and SAP Business by Design have been created. Some vendors (Acumatica, NetSuite, Plex) started with web-based architecture [21]. NetSuite and Plex are also pure SaaS vendors [22].

Most legacy ERP vendors are also Cloud ERP vendors, but they offer solutions as an on-premise ERP and a SaaS ERP other than legacy ERP software in the cloud. For example SAP [23] for small or mid-size enterprise offers SAP Business One or SAP All-in-One as an on-premise solution and SAP Business by Design as an on-demand solution. The SAP All-in-One can be also outsourced but cannot be used in the SaaS model. There are many examples of Cloud ERP in the world [22], [24]:

- Acumatica Cloud ERP software,
- Epicor Express on demand ERP,
- Infor ERP SyteLine in the Cloud,
- NetSuite Cloud ERP,
- Plex Online Cloud ERP,
- Sage Accpac Online,
- SAP Business by Design.

Epicor Express [25] was the Product of the Year 2010 for Customer Inter@ction Solution. It is the next generation of ERP software built on the concepts of SOA and BPM and which provides broad functionality as Web services grouped in the following modules [25]: financial management, material management, product management, production management, customer relationship management, business intelligence. Furthermore Epicor offer Web 2.0 and mobile solutions in order to increase productivity and to facilitate collaboration between users. The core of Epicor is on-demand business architecture which is a platform for integration of different applications. This SaaS ERP solution has been developed for small and mid-size enterprises to reduce complexity and decrease on-going operational costs [25].

The pricing model of EaaS is based on a monthly subscription fee, which in the case of Plex Online Cloud ERP includes uptime and performance, security, data transfer and storage plus intermediate "availability of enhancements on an opt-in basis" [18]. The provider deals with server administration, maintenance, implementation, problem solving and user training. The monthly fee for EaaS depends on the number of users and modules. The services in a cloud model do not require capital expense and are cheaper than hosting.

The vendor's choice of Cloud ERP is a very important business decision. The most important criteria for vendor selection are these [26]: support of changes in business processes, flexibility of usage model and security and service level agreements (SLA). A little less importance for the potential customer are [26]: scalability, training and support, integration with other applications, archiving and data recovery. Before implementing a Cloud ERP, the organization must prepare a security policy.

3 Benefits and Challenges of Cloud ERP in SaaS Model

Benefits vary when organizations use cloud only or cloud with web-based software. The benefits of cloud only are [8], [21]: scalability, pay for use and reduced IT expenditure. When a cloud utilizes web-based software there are additional benefits [8], [21]: SaaS ready, no client software, real-time data, faster to implement, simplified remote access (not VPN), easy to maintain and cross platform compatibility. The benefits of SaaS are [8]: lower up-front fees, rapid installation and less maintenance hassle. When a web-based ERP system is running as a SaaS in the cloud, the organization gains the benefits of three models with one deployment [8].

Cloud ERP providers can offer three types of services with different benefits and challenges for the organization (Table 2). In the case of the ERP system, the most important is flexibility because business needs are changing.

Table 2. Benefits and challenges of three types of Cloud ERP (Source: based on [20])

Type of Cloud ERP	Benefits	Challenges
SaaS using a Cloud Infrastructure	Maximizes flexibility to switch providers or move on-premise	Hosted service with lower pricing structure
SaaS using a Cloud Platform	Mix of flexibility and savings	Coordination challenges – vendor manages the application while service provider manages infrastructure
SaaS using a Cloud Application	Maximizes efficiencies for universal applications	Vendor lock-in; customer does not have option to move application to a different provider

The main benefits of Cloud ERP, according to an Institute of Management Accountants (IMA) survey, one of the largest cloud based ERP surveys (around 800 responses) [27], are (Fig. 3): lower total cost of ownership TCO (30%), data access anytime and anywhere (28%), streamlines business process (21%), easy upgrades (9%), lower capacity requirements (7%) and speed of deployment (5%). The cloud concept allows a reduction in IT outlay while at the same time increasing its efficiency. The user needs only a personal computer and secure Internet access.

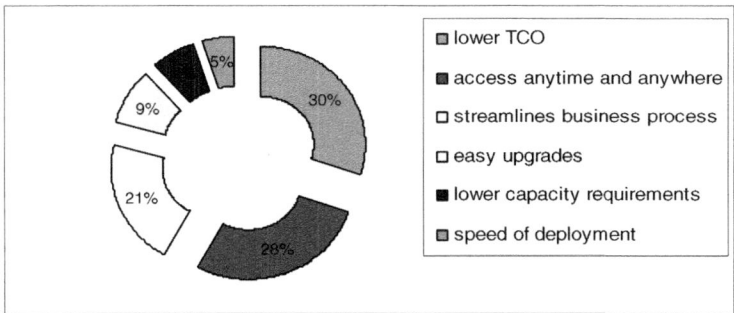

Fig. 3. Benefits of Cloud ERP (Source: drawn up on the basis of [27])

The increase of interest in cloud computing is connected with innovations in virtualization and distributed computing and improvements in data security and high-speed Internet. The main problems of cloud based ERP according to IMA [27] are (Fig. 4): security (35%), customization (18%), reliability versus an in-house (14%), ownership of data (12%), no substantial concerns (9%), maturity versus on-premise (8%) and ownership of application (4%).

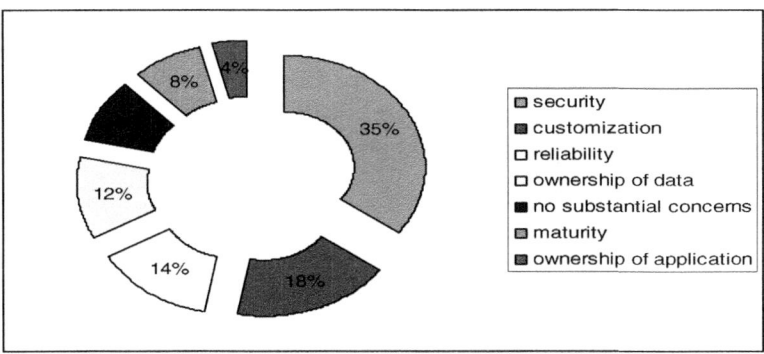

Fig. 4. Problems of Cloud ERP (Source: drawn up on the basis of [27])

The challenges of the cloud concept are concerned with security risks and compliance laws. The most important risks are connected with the change in management method and data security issues and also the transfer of data to and from the cloud.

The transfer process should be encrypted and user and data authentication should be ensured. The faults of cloud computing are the risk for businesses in case of an interruption in service delivery and the loss of control over valuable data, as well the risk of data leakage. The cloud stores critical business documents which should be kept in a safe location in order to guarantee data privacy of sensitive information. In the cloud, the strength of Internet connection is very important. The access and response times should especially be taken into account. Data transfer over the Internet can cause some delays. This means an application in the cloud model can work more slowly than in the company data centre. Moreover terabytes of data are unsuitable for transfer over the Internet.

An organization considering a Cloud ERP should analyze the contract to avoid hidden costs and potential problems [28]. The cloud contract has to be in accordance with international and Polish law regarding, for example: the accountancy act, personal data security act, database security act and act of service provided in an electronic fashion. According to the accountancy act, account books have to be kept in the company or, if outside the company, then on Poland soil. In the case of computer book-keeping, it is necessary to use software and organizational tools for data security. The CSP provider does not bear responsibility for consequential loss or irrecoverable benefits on the part of the customer which result from the use of services provided. In the global economy, take-over or merger of companies is always an option, so the client should make sure that the contract with the provider contains an assignment clause which guarantees continuity of service provision under such conditions. The SAP on demand standard contract, for example, allows transference of contract only in the case of SAP taking over another company [28].

The EaaS model has altered the relationships between business and IT managers. The Cloud ERP does not eliminate the need for IT department staff, because users still require access to the Internet and application configuration. The cloud concept allows IT managers to spend more time solving business problems and analyzing business data and technical IT staff can spend less time managing servers. "This allows IT employees to shift from being an unwanted expense to become an integral part of company profitability" [29]. With the Cloud ERP, operational IT tasks are outsourced but business expertise in-house is still a requirement [30]. The organization will need technical experts with a business background.

4 Conclusions

The terms SaaS and cloud computing should not be confused. Software as a service uses a cloud, which provides the computing power to run the solution, but it is not a cloud itself. It is, however, the most popular delivery model of cloud computing. A cloud is a less flexible but most cost-efficient method of implementing on-premise or on-demand ERP solutions. There is also the difference between IT outsourcing and cloud computing. In a public cloud, the customer does not know where the SaaS ERP system is located and the contract with the outsourcer is more long term and complicated than with cloud computing.

The most important values of the Cloud ERP model are reduction of hardware and license costs, scalability and manageability. The challenges of Cloud ERP are security, flexibility, integrity of the provider and ability to move to other provider.

Nowadays, the customer wishing to rapidly increase their business capabilities can combine on-line services with an on-premise or an on-demand ERP system. The concept of cloud computing will mature when it is able offer configurable modules which could be easily integrated with company specific software (in-house, on-premise or on-demand).

References

1. Laudon, K.C., Loudon, J.P.: Essentials of Management Information Systems, 8th edn. Pearson, Upper Saddle River (2009)
2. Rhoton, J. (ed.): Cloud Computing Explained: Enterprise Implementation Handbook, 2nd edn. Recursive Limited (2010)
3. Mather, T., Kumaraswamy, S., Latif, S.: Cloud Security and Privacy: An Enterprise Perspective on Risks and Compliance (Theory in Practice). O'Reilly, Sebastopol (2009)
4. National Institute of Standards and Technology, http://csrs.nist.gov
5. Piątkowski, Ł.: In the Clouds with Head (in Polish), http://www.wyzwaniafirm.pl
6. Sosinski, B.: Cloud Computing Bible. Wiley & Sons, Hoboken (2011)
7. Żur, Ł.: Cloud Computing, or Business in the Clouds (in Polish), http://nowetechnologie.comarch.com
8. Web-based, SaaS, and Cloud ERP benefits, http://erpcloudnews.com
9. Pawłowicz, W.: Two Small "a" and Great Changes. Networld (11) (2010) (in Polish), http://www.networld.pl
10. Solutions in IT Security Systems based on Cloud Computing Model (in Polish), http://www.storio.pl
11. Bradford, M.: Modern ERP - Select, Implement and Use Today Advanced Business Systems, 2nd edn. (2010), http://lulu.com
12. Lenart, A.: ERP Systems. In: Wrycza, S. (ed.) Business Informatics. PWE, Warszawa (2010) (in Polish)
13. ERP Software Predictions – (2011), http://erpcloudnews.com
14. ERP Software Cost Comparison: On-Premise, SaaS, and Hosted, http://erpcloudnews.com
15. Public Cloud (and) (or) Private Cloud, http://erpcloudnews.com
16. ERP Deployment Model, http://whatiserp.net
17. SaaS ERP: Trends & Observations for SME 2010. Aberdeen Group (June 2010), http://www.infor.com
18. What is Cloud ERP, http://www.plex.com
19. Confusion between Cloud ERP and SaaS ERP software, http://erpcloudnews.com
20. Different Types of Cloud ERP, http://erpcloudnews.com
21. Cloud ERP and Web-Based Software, http://erpcloudnews.com
22. Singleton, D.: The Cloud ERP Short List for Manufacturers. Cloud Computing Journal (2011), http://cloudcomputing.sys-con.com
23. ERP for SME, http://www.sap.com
24. Wailgum, T.: The big players in SaaS ERP. CIO (March 2010), http://www.cio.com

25. Epicor Solution Overview, `http://www.epicor.com`
26. What Criteria Should We Use For Selection of SaaS Provider (in Polish),
    `http://decyzje-IT.pl`
27. Turner, P.: The IMA Survey Results Are in - What the Cloud Means to Finance,
    `http://www.netsuiteblogs.com`
28. Cieśla, S., Helbing, T.: Legal aspects of Cloud computing (in Polish),
    `http://decyzje-IT.pl`
29. Using the Cloud To Weatherproof your Financials, `http://erpcloudnews.com`
30. One CIO's View of Cloud Computing and ERP Software,
    `http://erpcloudnews.com`

Building Project Teams in Enterprise System Adoption: The Need for the Incorporation of the Project Type

Piotr Soja

Cracow University of Economics
Department of Computer Science, Rakowicka 27, 31-510 Krakow, Poland
eisoja@cyf-kr.edu.pl

Abstract. This study's goal is to investigate and better understand project teams' building in enterprise system (ES) adoption. The analysis builds on two-phased research conducted among two groups of ES adopters. The investigated issues, revealed during the first, exploratory phase of the research, include the steering committee composition, the project team composition, and the involvement of the system provider's representatives. Next, their influence on ES adoption success is investigated during the second phase. In doing so, the analysis examines the impact of the project type defined by four criteria: adoption scope, inter-organizational focus, company size, and project duration time. Recommendations regarding project team building and future research conclude the paper.

Keywords: enterprise system, ERP, adoption, project team building, success.

1 Introduction

Enterprise systems (ES) are complex application software packages that contain mechanisms supporting the management of the whole enterprise and integrate all areas of its functioning [1]. ES evolved from Manufacturing Resource Planning (MRPII) and Enterprise Resource Planning (ERP) systems and started to include support for front-office and even inter-organizational activities including supply chain management (SCM), customer resource management (SCM), and sales force [10]. The adoption of an ES is a multi-staged project which involves different people and teamwork is an important implementation issue of enterprise systems.

During ES adoption, project participants are usually organized around two main bodies: the steering committee and the project team. The steering committee typically include senior management from different corporate functions and is usually involved in system selection, monitoring during implementation, and management of outside consultants [6]. The project team includes highly respected individuals from each function who should be entrusted with decision making responsibility [8]. These two main bodies are headed by the chief of the steering committee and the project manager, respectively. Additionally, during ES adoption, it is important to involve in the project teams not only people representing various functional areas of the organization, but also the supplier's consultants [9].

The goal of this paper is to examine and better understand the issues connected with project teams organization in ES implementation and their influence on ES

S. Wrycza (Ed.): SIGSAND/PLAIS 2011, LNBIP 93, pp. 51–65, 2011.

adoption success. In doing so, considering the huge diversity of ES adoptions, this study investigates project team-related issues depending on project type. The analyzed perspectives include ES adoption scope, inter- vs. intra-organizational focus, company size, and project duration time. The issues investigated concentrate on the project team composition, the project manager, the steering committee arrangement, and the provider's representatives involvement.

2 Research Methodology

2.1 Research Design

The research question employed by this study can be formulated as follows:

- How to organize project teams in order to achieve success in enterprise system adoption?

In order to answer the research question, this research employs a multi-method approach which consists of two empirical sub-studies conducted among ES adopters. The research phases and involved steps are illustrated in Fig. 1.

The purpose of the first phase was to explore issues describing project teams' building during ES adoption project and to preliminary reveal their impact on the ES adoption success. The first sub-study had a qualitative nature and was based on a field study conducted among 41 ES adopters. The gathered data and its analysis allowed us to discover issues connected with various people and groups playing significant roles during ES adoption project.

The results of the first study and revealed issues served as a starting point for the second study, which had mainly quantitative character. The research questionnaire has been worked out with the purpose of gathering data describing the discovered project team-related issues. Next, using the elaborated research instrument, a field study has been conducted during which 140 ES adopters have been inquired.

The data gathered during the Phase II and its further analysis allowed us to define independent variables describing various aspects of project teams composition. In order to discover the influence of these variables on the ES adoption outcome, ES adoption success measure has been adopted as a depended variable. The analysis also seeks to discover the role of project type in project teams building and in doing so it employs several intervening variables.

2.2 Issues Investigated

The data gathered during the first exploratory stage of the research and its analysis revealed a number of issues connected with project teams building in the context of ES adoption. The following paragraphs describe shortly the issues discovered and their influence on ES adoption successfulness. The revealed concepts are connected with the steering committee composition, the characteristics of the chief of the steering committee, the project team composition, and the characteristics of the project manager. More details can be found in [4].

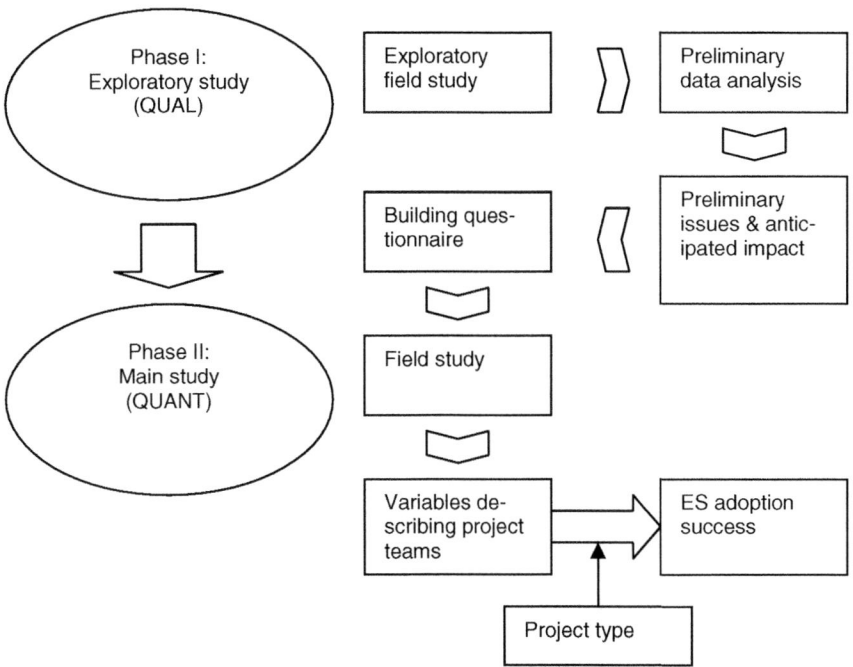

Fig. 1. Research phases and steps

The steering committee. The steering committee-related issues are connected with the representation of IT department in the committee and the presence of the system provider's representatives. The preliminary results suggest that projects that appointed a steering committee achieved a slightly higher level of success metrics. However, no statistical significance was discovered. With respect to an IT/IS person's membership in the steering committee, the preliminary results show that projects where IT/IS people were not present in a steering committee achieved a slightly higher level of all success metrics.

The chief of the steering committee. The chief of the steering committee was investigated with respect to his/her organizational position. The initial results suggest that the most successful were projects employing top management representatives as the head of a steering committee. Additional issue suggested by the first study involves the organizational area of the head of the steering committee, however, no initial impact of this issue was researched.

The project team. The project team composition-related issues include team completeness, the position of an IT person within the team, and the involvement of the system provider's representative. The preliminary findings regarding team completeness, defined as the involvement of people responsible for each system module introduced, are mixed. The results surprisingly suggest that success metrics decrease when team completeness increases and, on the other hand, imply that the projects involving more people tend to achieve somewhat higher levels of success measures. The results also suggest that the higher the position of an IT person within the project team, the better results achieved. Finally, the presence of the system supplier representative in the project team had no influence on ES success measures.

The project manager. The project manager-related issues include his/her organizational position, IT background and functional department. The preliminary findings suggest that adoptions where functional managers played the role of the project manager revealed the highest level of user satisfaction. Next, on average, ES implementations employing project managers from outside the IT/IS department achieved higher levels of success metrics. Also, in the case of ES adoptions managed by IT people, a tendency can be observed that projects led by the IT people of lowest organizational rank (i.e. specialists) achieved the lowest levels of success metrics. Finally, the examination of the project manager's functional department showed mixed results.

2.3 Questionnaire Items and Variables Definition

The outcome of the first research phase was used during the definition of the Phase II's questionnaire items and during the consecutive variables' definition. The elaborated issues are described in Table 1 which contains variables names and their definitions. All variables are dichotomous and may accept values 0 or 1.

Table 1. Project Teams-related Variables

Variable name	Issue/Variable description
SteerComm	Was a steering committee appointed?
SteerCommChiefOperPos	Organizational position of the chief of the steering committee: Operational?
SteerCommChiefManagPos	Organizational position of the chief of the steering committee: Managerial?
SteerCommChiefDept (Top)	Department of the chief of the steering committee: Top management?
SteerCommChiefDept (Finance)	Department of the chief of the steering committee: Finance?
SteerCommChiefDept (IT)	Department of the chief of the steering committee: IT?
SteerCommITOperationalPos	Organizational position of an IT representative in the steering committee: Operational?
SteerCommITManagerialPos	Organizational position of an IT representative in the steering committee: Managerial?
SteerCommProviderOperPos	Organizational position of the provider's representative in the steering committee: Operational?
SteerCommProviderManagPos	Organizational position of the provider's representative in the steering committee: Managerial?
ProjectManagerOperationalPos	Organizational position of the project manager: Operational?
ProjectManagerManagerialPos	Organizational position of the project manager: Managerial?
ProjectManagerDept (Top)	Department of the project manager: Top management?
ProjectManagerDept (Finance)	Department of the project manager: Finance?
ProjectManagerDept (IT)	Department of the project manager: IT?
ProjectTeamITOperationalPos	Organizational position of an IT representative in the project team: Operational?
ProjectTeamITManagerialPos	Organizational position of an IT representative in the project team: Managerial?
ProjectTeamProviderOperPos	Organizational position of the provider's representative in the project team: Operational?
ProjectTeamProviderManagPos	Organizational position of the provider's representative in the project team: Managerial?
ProjectTeamAreAllDepartments	Were all departments represented in the project team?

2.4 Research Model

The main goal of this study is to investigate the influence of the project team-related issues on ES adoption success, which is depicted in the final research model in Fig. 2. The project team-related elements are independent variables while ES adoption success is the dependent variable and is defined as user satisfaction [3, 5]. Specifically, this study employs a four-item instrument covering the following issues: information processing needs, system efficiency, system effectiveness, and satisfaction from the system [2]. All items are measured on a 7 point Likert scale from 1 to 7.

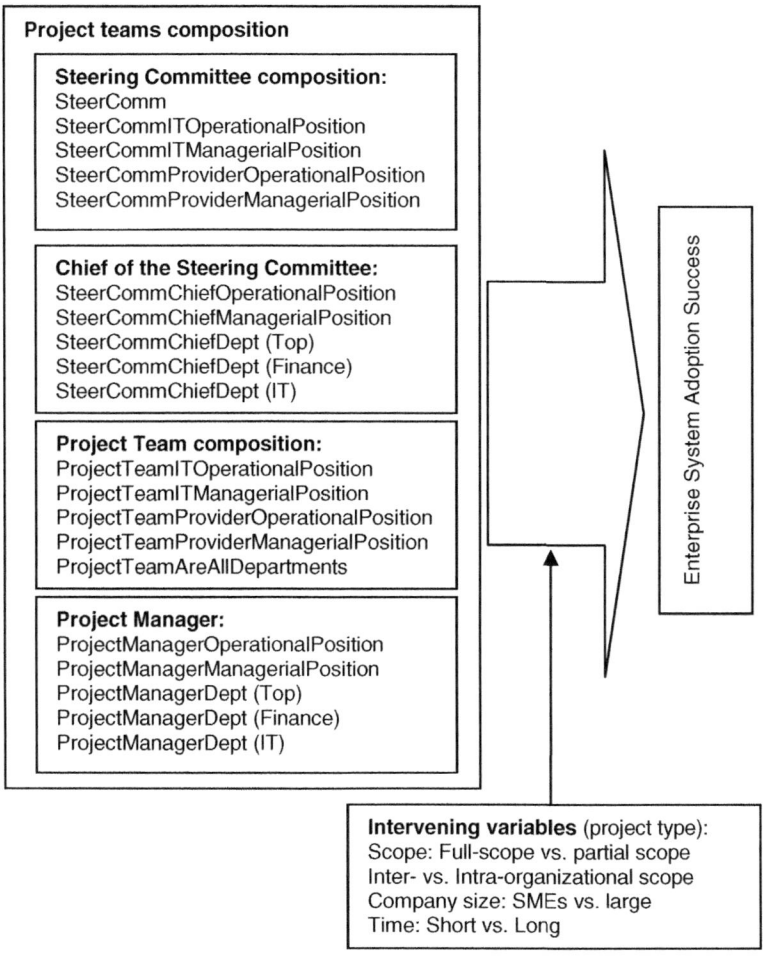

Fig. 2. Final research model

This study also seeks to discover the impact of several project characteristics, defined by the intervening variables in Fig. 2, on the project teams building. In doing so, the analysis divides the investigated projects into pairs of groups taking into consideration the variables and their median values. In consequence, the projects were divided into full- and partial-scope projects, inter- and intra-organizational adoptions, projects conducted in SMEs and large companies, and adoptions with short and long duration time.

The influence relationship has been investigated using multiple regression techniques. Also, a stepwise regression has been employed in order to discover an optimal subset of variables [11 p.437]. Particularly, both approaches to stepwise regression have been used: backward elimination and forward selection. The calculations have been performed for the whole research sample and for all subsets.

3   Results

The following section shows the results of multiple regression calculations for all considered groups of projects. The number of elements in a group (row "N"), the model's statistical significance (row "p value <"), percentage of the variance explained (row "R2") are displayed for all types of calculated regression, which are arranged in columns. An empty column indicates that the regression calculations were not possible.

3.1   All Projects

The results of calculations for all researched projects (shown in Table 2) illustrate that the general regression model is not significant. Stepwise regression resulted in significant

Table 2. Multiple and Hierarchical Regression Coefficients for All Companies

Variable	Multiple regress.	Stepwise regression	
		backward eliminat.	forward selection
Project Type	All	All	All
N	140	140	140
p value <	0.29	0.00	0.01
R2	0.17	0.09	0.15
Regression coefficients:			
SteerComm	0.13		
SteerCommChiefOperPos	0.34		0.20
SteerCommChiefManagPos	0.12		
SteerCommChiefDept (Top)	0.34	0.34^^	0.36^
SteerCommChiefDept (Finance)	0.08		
SteerCommChiefDept (IT)	0.03		
SteerCommITOperationalPos	-0.01		
SteerCommITManagerialPos	-0.19		
SteerCommProviderOperPos	-0.30		
SteerCommProviderManagPos	-0.51^	-0.44*	-0.40*
ProjectManagerOperationalPos	-0.38	-0.33*	
ProjectManagerManagerialPos	-0.04		0.31^
ProjectManagerDept (Top)	0.16		
ProjectManagerDept (Finance)	0.43		0.36
ProjectManagerDept (IT)	-0.04		
ProjectTeamITOperationalPos	0.37		0.34
ProjectTeamITManagerialPos	0.57*		0.47*
ProjectTeamProviderOperPos	-0.04		
ProjectTeamProviderManagPos	-0.31		-0.23
ProjectTeamAreAllDepartments	0.39*	0.34*	0.30^

Notes: *p<0.05, **p<0.01, ^p<0.06, ^^p<0.08

models explaining a rather low percentage of variance and indicates the significance of several variables. The results reveal the positive influence of involving the company's all departments in ES adoption and suggest the positive impact of involving IT managers in the project team. Additionally, the calculations imply the positive influence of the chief of the steering committee coming from the company's top management, however, in this case there is a border significance. The results interestingly illustrate negative influence of several issues. The first issue refers to the presence of the provider's representative in the steering committee having a managerial organizational position. The second issue is connected with negative impact of low organizational position of the project manager. This is nicely supplemented with indications of positive influence of the project manager holding a managerial position.

3.2 Inter- vs. Intra-organizational

The calculations taking into consideration organizational scope indicate the paramount role of the steering committee in inter-organizational projects. For projects of this kind, the stepwise regression with forward selection resulted in a significant model explaining

Table 3. Multiple and Hierarchical Regression Coefficients by Organizational Scope

Variable	Multiple regress.	Stepwise regression		Multiple regress.	Stepwise regression	
		backward eliminat.	forward selection		backward eliminat.	forward selection
Project Type	Inter	Inter	Inter	Intra	Intra	Intra
N	48	48	48	92	92	92
p value <			0.00	0.70	0.05	0.04
R2			0.46	0.19	0.07	0.14
Regression coefficients:						
SteerComm			1.02**	-0.22		
SteerCommChiefOperPos				0.40		
SteerCommChiefManagPos				0.05		-0.30
SteerCommChiefDept (Top)				0.42		0.41
SteerCommChiefDept (Finance)				0.20		
SteerCommChiefDept (IT)				-0.15		
SteerCommITOperationalPos				-0.24		
SteerCommITManagerialPos				-0.31		
SteerCommProviderOperPos				-0.19		
SteerCommProviderManagPos			-0.42^^	-0.41	-0.39^^	-0.43^^
ProjectManagerOperationalPos				-0.34		
ProjectManagerManagerialPos			0.41	0.06		0.32
ProjectManagerDept (Top)				0.20		
ProjectManagerDept (Finance)			0.54	0.40		0.28
ProjectManagerDept (IT)				-0.11		
ProjectTeamITOperationalPos			0.45*	0.29		
ProjectTeamITManagerialPos				0.70^	0.39^^	0.45*
ProjectTeamProviderOperPos				0.09		
ProjectTeamProviderManagPos				-0.08		
ProjectTeamAreAllDepartments				0.17		

Notes: *p<0.05, **p<0.01, ^p<0.06, ^^p<0.08; empty column: bad regression matrix
Inter-organizational: CRM or SCM adopted; Intra- organizational: no CRM nor SCM adopted

46% of variance. Another issue positively influencing inter-organizational adoptions is connected with the involvement of an IT representative in the project team at the operational level. Finally, the results illustrate that the presence of the provider's representative in the steering committee holding managerial position may have negative influence on the project success.

The calculations performed for intra-organizational projects agree with respect to this issue, however, we should emphasize that the percentage of variance explained by the regression model is much lower for intra-organizational adoptions. Overall, adoption projects of this kind are positively influenced by the presence of IT representatives holding managerial positions in the project team. The results also suggest positive influence of some factors regardless of organizational scope. This refers to the presence of the project manger holding managerial position and representing the financial department. Nonetheless, the results are not statistically significant for these issues.

3.3 Full vs. Partial Scope

Incorporating the adoption project scope, understood in terms of a number of implemented modules, we reveal that the adopted model does not fit full-scope adoptions – all calculations yield either bad regression matrix or lack of statistical significance (Table 4).

Table 4. Multiple and Hierarchical Regression Coefficients by Project Scope

Variable	Multiple regress.	Stepwise regression		Multiple regress.	Stepwise regression	
		backward eliminat.	forward selection		backward eliminat.	forward selection
Project Type	Partial	Partial	Partial	Full	Full	Full
N	73	73	73	67	67	67
p value <	0.00	0.00	0.00			0.22
R2	0.54	0.43	0.50			0.07
Regression coefficients:						
SteerComm	1.05					
SteerCommChiefOperPos	-0.93					
SteerCommChiefManagPos	-1.15					
SteerCommChiefDept (Top)	0.57*	0.52*	0.65**			
SteerCommChiefDept (Finance)	0.33		0.33			
SteerCommChiefDept (IT)	-0.21					
SteerCommITOperationalPos	0.44		0.36^^			
SteerCommITManagerialPos	0.12					
SteerCommProviderOperPos	-0.36		-0.27			
SteerCommProviderManagPos	-0.63*	-0.57**	-0.70**			
ProjectManagerOperationalPos	-0.95	-0.52**				
ProjectManagerManagerialPos	-0.24		0.66**			
ProjectManagerDept (Top)	-0.39		-0.23			0.45
ProjectManagerDept (Finance)	-0.30					0.44
ProjectManagerDept (IT)	-0.50					
ProjectTeamITOperationalPos	0.53*	0.56**	0.41^			
ProjectTeamITManagerialPos	0.79**	0.83**	0.71**			
ProjectTeamProviderOperPos	0.06					
ProjectTeamProviderManagPos	-0.23					-0.30
ProjectTeamAreAllDepartments	0.65**	0.43**	0.55**			

Notes: *$p<0.05$, **$p<0.01$, ^$p<0.06$, ^^$p<0.08$; empty column: bad regression matrix;
Core modules: Finance, Purchasing, Inventory, Sales, Shop Floor Control, MRP Explosion, Production Planning; Partial scope: sum of core modules <= 3 (Median value); Full scope: sum of core modules > 3

However, on the other hand, the model reveals a reasonable fit for partial-scope projects with more than 40% of variance explained. There are a number of issues having strong positive influence on ES adoption success: involvement of company's all departments, chief of the steering committee representing company's top management, and presence of an IT manager in the project team. With respect to the latter, the results also suggest that involvement of anyone from the company's IT department is beneficial for the adoption project successfulness.

The results also illustrate clear negative influence of the presence of the provider's managers in the steering committee. Further, there are some interesting observations regarding the project manager's organizational position. Namely, the results suggest that it might be beneficial for the project to employ a project manager holding managerial position. Simultaneously, the calculations suggest that having the project manager who holds operational position within the company's organization might be detrimental for the project success.

3.4 SMEs vs. Large Companies

Considering the company size we deal with the similar situation as in the case of project scope – lack of model fit for one group (i.e. large companies), and a reasonable fit for the

Table 5. Multiple and Hierarchical Regression Coefficients by Company Size

Variable	Multiple regress.	Stepwise regression		Multiple regress.	Stepwise regression	
		backward eliminat.	forward selection		backward eliminat.	forward selection
Project Type	SME	SME	SME	Large	Large	Large
N	79	79	79	57	57	57
p value <	0.18	0.00	0.00			0.11
R2	0.33	0.21	0.32			0.13
Regression coefficients:						
SteerComm	-0.07		0.62*			
SteerCommChiefOperPos	0.97					
SteerCommChiefManagPos	0.73					-0.33
SteerCommChiefDept (Top)	0.35	0.52*	0.32			
SteerCommChiefDept (Finance)	0.32					
SteerCommChiefDept (IT)	0.11					
SteerCommITOperationalPos	-0.16					
SteerCommITManagerialPos	0.05					
SteerCommProviderOperPos	-1.15**	-0.49*	-0.98**			0.59*
SteerCommProviderManagPos	-1.19**	-0.68**	-1.12**			
ProjectManagerOperationalPos	-1.03	-0.55*	-1.19^			
ProjectManagerManagerialPos	-0.43		-0.65			
ProjectManagerDept (Top)	0.36		0.33			
ProjectManagerDept (Finance)	0.45		0.47			
ProjectManagerDept (IT)	-0.04					-0.31
ProjectTeamITOperationalPos	0.50		0.40			
ProjectTeamITManagerialPos	0.36		0.32			
ProjectTeamProviderOperPos	0.01					
ProjectTeamProviderManagPos	0.03					
ProjectTeamAreAllDepartments	0.57*	0.53*	0.58**			0.48

Notes: *p<0.05, **p<0.01, ^p<0.06, ^^p<0.08; empty column: bad regression matrix;
SME: number of employees <= 250; Large: number of employees > 250

other group of projects (i.e. SMEs). For projects conducted in SMEs, the results illustrate a positive influence of involvement of company's all departments and suggest a positive role of the chief of the steering committee representing company's top management. Simultaneously, the calculations indicate the clear negative influence of the presence of the provider's representatives in the steering committee, regardless of their organizational positions. Similarly, the results suggest the negative influence of the project manager having a lower organizational position.

3.5 Short vs. Long Projects

The calculations performed for short and long projects illustrate that the project duration time is a characteristic differentiating significantly ES adoption considerations (Table 6). Particularly, both groups of projects agree upon the positive influence of appointing the steering committee. Apart from that, there is a number of issues influencing only one group of projects or having diverse influence. The latter applies to appointing the chief of the steering committee from the financial department, which is beneficial for short projects and seems to be detrimental for long projects.

Table 6. Multiple and Hierarchical Regression Coefficients by Project Duration Time

Variable	Multiple regress.	Stepwise regression		Multiple regress.	Stepwise regression	
		backward eliminat.	forward selection		backward eliminat.	forward selection
Project Type	Short	Short	Short	Long	Long	Long
N	58	58	58	74	74	74
p value <			0.00	0.69	0.04	0.03
R2			0.38	0.25	0.11	0.20
Regression coefficients:						
SteerComm			0.67*	-0.19	0.55^^	
SteerCommChiefOperPos				0.55		0.66*
SteerCommChiefManagPos				-0.12		
SteerCommChiefDept (Top)				0.11		
SteerCommChiefDept (Finance)			0.70*	-0.44	-0.51^^	-0.51^^
SteerCommChiefDept (IT)				0.16		
SteerCommITOperationalPos				0.30		
SteerCommITManagerialPos			-0.27	-0.13		
SteerCommProviderOperPos			-0.83**	0.22		
SteerCommProviderManagPos			-0.76*	0.14		
ProjectManagerOperationalPos			-0.70**	0.45		
ProjectManagerManagerialPos				0.25		
ProjectManagerDept (Top)				0.37		0.33
ProjectManagerDept (Finance)				0.53		0.50
ProjectManagerDept (IT)				-0.07		
ProjectTeamITOperationalPos			0.17	0.33		0.51
ProjectTeamITManagerialPos				0.83	0.38^^	0.88*
ProjectTeamProviderOperPos				-0.07		
ProjectTeamProviderManagPos				-0.52		-0.41
ProjectTeamAreAllDepartments			0.60**	0.22		

Notes: *p<0.05, **p<0.01, ^p<0.06, ^^p<0.08; empty column: bad regression matrix
Short: duration time < 7 months (Median value); Long: duration time >=7 months

Short projects are positively influenced by involvement of company's all departments while long projects benefit from involving an IT manager in the project team and may be positively influenced by the chief of the steering committee who holds a lower organizational position. In the case of short projects the results indicate a strong negative influence of the presence of the provider's representatives in the steering committee regardless of their organizational position. Further, the calculations reveal a strong negative influence of the project manager who holds a low organizational position.

4   Discussion of Findings

4.1   The Influence of Adoption Participants by Project Type

The influence of various project participants on ES adoption success depending on project type was summarized in Table 7. The positive influence of a given variable has been marked by a plus sign, negative influence has been denoted by a minus sign. Two types of stepwise regression have been taken into account: backward elimination or forward selection.

Table 7. Influence of Project Team-related Variables by Project Type

Variable	All	Part.	Full	Inter	Intra	SME	Large	Short	Long
SteerComm				+		+		+	+
SteerCommChiefOperPos									+
SteerCommChiefManagPos									
SteerCommChiefDept (Top)	+ +	+ +							
SteerCommChiefDept (Finance)								+	- -
SteerCommChiefDept (IT)									
SteerCommITOperationalPos		+							
SteerCommITManagerialPos									
SteerCommProviderOperPos						- -			-
SteerCommProviderManagPos	-	- -		-	- -	- -			-
ProjectManagerOperationalPos	-	-				- -			
ProjectManagerManagerialPos	+	+							
ProjectManagerDept (Top)									
ProjectManagerDept (Finance)									
ProjectManagerDept (IT)									
ProjectTeamITOperationalPos		+ +		+					
ProjectTeamITManagerialPos	+	+ +			+ +				+ +
ProjectTeamProviderOperPos									
ProjectTeamProviderManagPos									
ProjectTeamAreAllDepartments	+ +	+ +				+ +		+	

Note: a sign (+ or –) denotes that a variable was significant or border significant in stepwise regression (backward elimination or forward selection)

In general, regardless of the project type, the results suggest that appointing the chief of the steering committee representing company's top management is beneficial for the adoption success. Simultaneously, with respect to the steering committee composition, the findings interestingly illustrate that the presence of the provider's representative holding managerial position has negative influence on the project success. The findings related with the project manager illustrate that appointing the project manager having low organizational position is negative for the project, while appointing a person holding managerial position has positive influence on the project success. As regards the project team composition, there is a positive influence of involving representatives of company's all departments and IT managers in the project team.

The impact of the issues influencing ES adoption project success regardless of project type varies depending on an analysis perspective. Some issues lose their significance and, on the other hand, sometimes new influential elements appear. In the case of partial-scope adoptions, all variables having influence for all companies also have impact on these projects. Additionally, there are two new influential variables describing the positive impact of the presence of IT specialists in the steering committee and in the project team. This implies that for partial adoptions the IT representative's organizational position does not matter: anyone from the company's IT department involved in the project team would benefit the project.

The implementation of inter-organizational modules such as CRM or SCM does not change much the adoption considerations. In general, the positive role of appointing the steering committee appears and some shift in the required organizational position of an IT representative in the project team can be observed. Particularly, in inter-organizational projects the operational position is sufficient, while intra-organizational projects seem to require higher organizational position of an IT representative. Finally, regardless of organizational scope, the projects suffer from the presence of the higher-rank provider's representatives in the steering committee.

The smaller size of the company requires the appointment of the steering committee and also causes negative influence of the provider's representatives in the steering committee regardless of their organizational ranks. SMEs also follow the general rule connected with the negative impact of the project manager holding low organizational position. Projects conducted among smaller companies also benefit from the involvement of company's all departments.

Shorter projects experience all considerations typical of adoptions conducted among SMEs. Additionally, the results interestingly illustrate that it is beneficial for shorter projects to appoint a person from the financial department as the chief of the steering committee. However, this variable reveals negative influence for longer projects which illustrates potential shift in ES adoption considerations connected with the project duration time. Nevertheless, this is a preliminary finding and it requires further investigation. Increased project duration time causes the greater influence of IT department's representatives in the project team and also reveals the greater impact of the chief of the steering committee. Interestingly, the results suggest that for longer projects it is beneficial to employ an individual holding low organizational position.

4.2 Implications for Team Building

The results of the analysis illustrate several interesting implications regarding team building in the enterprise system adoption context. They illustrate some tendencies connected with the choice of the adoption participants and suggest whose stakeholders deserve more attention depending on project type. In general, the findings shed more light on several critical ES adoption considerations such as cooperation with the system provider, project management and involvement of IT people in the adoption.

The results imply that it is beneficial for the project success to appoint the steering committee, especially for shorter projects and those conducted in smaller companies, and also deliver several suggestions as regards the steering committee composition. These mainly refer to the characteristics of the chief of the steering committee. In general, regardless of the project type, it is beneficial to appoint the representative of company's top management in this role. Specifically, partial implementations should ensure this requirement. However, on the other hand, longer projects seem to require for this position an individual holding lower organizational rank, which might be explained by the probable difficulty of having a higher-up involved in this role for an extended period of time. The results also imply that the choice of this person should be made with care and should be accompanied by the careful examination of the project considerations, as the inappropriate business background might have a detrimental influence on the project success. Specifically, this applies to longer projects which may suffer from the appointment of the chief of the steering committee representing the company's financial department.

The results interestingly indicate that companies should avoid the presence of the provider's representatives in the steering committee. This implication specifically refers to the representatives holding higher organizational seniority. Moreover, shorter projects and those conducted in SMEs should avoid the presence of any representative of the provider in the steering committee. This issue is connected with the findings regarding the project team composition and lack of influence of the provider's representative presence for any project type. Another issue connected also with the project team composition concerns the involvement of IT people during the project. Specifically, there are no clear indications as regards the involvement of IT people in the steering committee, except for partial projects which should benefit from the presence of an IT specialist in the steering committee.

The findings regarding the project team imply that, in general, companies should involve an IT manager in the project team. This is especially important for projects having intra-organizational scope and taking place for an extended period of time. For partial scope projects and those having inter-organizational scope, an IT representative might also represent lower organizational seniority. In general, regardless of the project type, the project team should involve representatives of company's all affected departments. Specifically, projects having shorter duration time and conducted in smaller companies should benefit from this suggestion.

The results indicate that companies should appoint a project manager holding a managerial position within the company and avoid a person with low organizational seniority in this role. This rule applies to the projects of any type and should be especially followed by partial scope adoptions and those taking place in SMEs and having a short duration time. The practitioners should be aware that projects of these kinds might be negatively affected by the presence of the project manager holding low organizational position.

The research results imply that ES adopters should take responsibility for their adoption projects and should ensure adequate personnel for project management activities. People responsible for project management should be involved around two bodies: the steering committee and the project team. The findings illustrate that the level of responsibility of these two groups may vary depending on the project type. Specifically, among shorter projects responsibility for the project successfulness shifts to the steering committee and a greater role of management personnel can be observed. In general, this study's findings seem to confirm the notion that the most effective way of using the provider's resources is to bring them in to work under in-company direction and control [12].

This study sheds new light on the cooperation between the adopter and the provider and suggest that the formal solutions consisting in involving the provider's representatives in project management bodies are not beneficial for the project success. What is worse, the findings suggest that such solutions can be even detrimental for the project. This might suggest that, with respect to the good cooperation between the adopter and the provider, the problem is to establish informal contacts and good rapport between the representatives of these two main parties involved in the project [7].

The main limitation of this study is connected with a lack of the model's significance for the full-scope projects and those conducted among large companies. This illustrates some avenues for future research which may focus on the refinement of the model and introducing new issues describing the optimal project team composition in the case of full-fledged ES adoptions. Also, the model proposed by this study might be re-tested in different organizational settings using an adequately large research sample.

5   Conclusion

This study examined issues connected with project teams' composition and their influence on enterprise system adoption success. The employed multi-study approach building on two field studies allowed us to reveal the influence of project team-related issues on ES adoption success and to investigate the role of the project type in this relation. The results indicate some universal rules that should be followed regardless of the project type, such as involvement of company's management and IT personnel; however, the findings also indicate that project team-related issues should be analyzed taking into consideration project scope, company size, and project duration time. The outcome of the research should be valuable for the practitioners as it suggests several rules that could be helpful in project team building and working out the effective cooperation with the system and implementation services provider. The directions of future research suggested by this study mainly refer to the refinement of the model with the purpose of incorporating full-fledged ES adoptions, i.e. those conducted in large companies and introducing an ES within its full-scope.

References

1. Davenport, T.H.: Putting the Enterprise into the Enterprise System. Harvard Business Review 76(4), 121–131 (1998)
2. Seddon, P.B., Kiew, M.-Y.: A Partial Test and Development of the DeLone and McLean Model of IS Success. In: Proceedings of the International Conference on Information Systems, Vancouver, Canada, pp. 99–110 (1994)

3. Sedera, D., Tan, F.T.C.: User Satisfaction: An Overarching Measure of Enterprise System Success. In: Proceedings of Pacific Asia Conference on Information Systems, pp. 963–976 (2005)
4. Soja, P.: Investigating the Impact of Project Team Composition in Enterprise System Implementation: an Exploratory Study. In: Proceedings of the 12th Americas Conference on Information Systems, Acapulco, Mexico, pp. 2457–2466 (2006)
5. Somers, T.M., Nelson, K., Karimi, J.: Confirmatory Factor Analysis of the End-User Computing Satisfaction Instrument: Replication within an ERP Domain. Decision Sciences 34(3), 595–621 (2003)
6. Somers, T., Nelson, K.: A taxonomy of players and activities across the ERP project life cycle. Information & Management 41, 257–278 (2004)
7. Themistocleous, M., Soja, P., Cunha, P.R.: The Same, but Different: Enterprise Systems Adoption Lifecycles in Transition Economies. Information Systems Management 28, 223–239 (2011)
8. Umble, E.J., Umble, M.M.: Avoiding ERP implementation failure. Industrial Management, 25–33 (January/February 2002)
9. Volkoff, O., Sawyer, S.: ERP Implementation Teams, Consultants, and Information Sharing. In: Proceedings of the Americas Conference on Information Systems, pp. 1043–1045 (2001)
10. Volkoff, O., Strong, D.M., Elmes, M.: Understanding enterprise systems-enabled integration. European Journal of Information Systems 14, 110–120 (2005)
11. Walpole, R.E., Myers, R.H., Myers, S.L.: Probability and Statistics for Engineers and Scientists, 6th edn. Prentice Hall International, Inc., Upper Saddle River (1998)
12. Willcocks, L., Sykes, R.: The role of the CIO and IT function in ERP. Communications of the ACM 43(4), 32–38 (2000)

Model of Information Systems' Selection for the Company Management

Iryna Zolotaryova and Anna Khodyrevska

Kharkiv National University of Economics,
Information Systems dept., Lenina ave, 9A, 61001, Kharkiv, Ukraine
{izolotaryova,khodyrevska}@gmail.com

Abstract. The article describes the features of an enterprise's business process management, modern techniques, their advantages and ways of realization. The analysis of the possible implementation problems of the techniques is shown and ways of overcoming. There were also identified some key factors for choosing a project for business process management. Here is also shown the diagram of business processes optimization based on evaluating the performance of Information System. The progress of the research process in the form of a questionnaire survey of companies' experts is also described, giving the developed form of the questionnaire and respondents' answers. The paper also describes the use of multiple regression models to research data.

Keywords: Information Technologies, Information System, enterprise architecture, ITIL, BPM.

1 Introduction

An enterprise architecture defines the type of Information system (IS) used. Information technologies have become profitable not only themselves, but they determine the competitiveness of the company, form and retain customers' and partners' base, ensure the return of company's investments. With the gradual transformation of information technologies (IT) in an essential attribute of an enterprise, IT and IS formed on their basis placed some new demands.

At the moment, seems impossible to consider methods of organization and implementation of IS without committing to the business requirements and its organizational structure. The choice of particular information technologies should be based on the business architecture and not on the trends in the IT field. The concept of business architecture is inextricably linked with the company's structure, its industry sector, occupational orientation and other characteristics [1]. Shared vision provided by an enterprise's architecture, gives an opportunity of a unified systems engineering, designed to meet the needs of the organization, as well as interaction and integration abilities.

During the design of enterprise's development strategies changes in the business enterprise architecture are detected to optimize business processes, which changes directly affects the change in IT architecture. The next step is to develop a transition plan from the current to the planned stage. The transition process is a step towards the

S. Wrycza (Ed.): SIGSAND/PLAIS 2011, LNBIP 93, pp. 66–77, 2011.

transformation of the enterprise, and its' ending means moving the organization to a new stage of development, beginning as a development strategy.

Type of models vary depending on the architectural methodology. For example, in accordance with the methodology of Casewise company, based on the Zachman Framework [2], for description of the enterprise architecture developers have to fill up 30 different types of related models, divided into five levels of abstraction. One solution is to use specialized software products focused on enterprise architecture modeling, allowing to link different levels of abstraction to each other, and having a single repository for data storage [1].

Software products used for modeling can be divided in the following areas, according to the approach suggested by IFEAD analysts (Institute for Enterprise Architecture Developments) [3]: Software Engineering, Service Oriented Architecture, Enterprise Architecture, Business and IT strategy, Enterprise and IT portfolio, Program Management, Governance, Risk, Compliancy.

Gartner analysts highlighted the following leading companies: IBM, Troux Technologies, IDS Scheer.

Modern companies often begin to develop an enterprise's architecture with the models. According to Gartner analysts, this is one of the most common mistakes, because simulation provides only specification and documentation of the information gathered in previous phases of the architectural process.

The specific method of enterprise architecture documenting and modeling has no decisive importance if the information is available and stored in visual form. Thus, we conclude that the use of complex architectural tools in the modeling process is not a prerequisite for obtaining a successful outcome.

2 Methodologies for Companies' IT-Management

The analysis of sources didn't identify a common methodology for assessing the effectiveness of IT-investment. Therefore, we have attempted to formalize this process by constructing a model of evaluating the effectiveness of IT-investment [3].

From the perspective of final business effects at the highest level of assessment of the potential economic benefits, significant areas that determine the economic efficiency of any investment are known as key factors of economic efficiency. These include: minimization of lost income or the creation of new sources, reduction of current production (operating) costs, reduction of administrative and management costs, minimizing taxes and other obligatory payments, reducing penalties and other non-operating expenses, reduced need for capital expenditure, increase of current assets turnover [2, 4].

The present indicator of total cash flow is usually expressed in remaining elements of the net income. This indicator is used as a generalized view of all significant effects, provided by the implementation of the investment project. After that any calculation of the effectiveness index (ROI, NPV, IRR, PP, etc.) can be provided for customer (investor).

The main difficulty in assessing of the impact of IT projects investments is the limited applicability of financial evaluation methods in relation to the need to consider non-financial benefits of IT projects [5, 6, 7].

Understanding the complexity of IT management Office of Government Commerce (the UK government department) has developed the document named IT Infrastructure Library (ITIL, 1989), which is a set of practical recommendations to improve service management by identifying the interaction between human capital, processes and technologies.

The main components of ITIL are: incident management, problem, change and configuration management. The use of the ITIL enables companies to improve utilization of existing resources, reduce the number of incidents caused by internal factors, to improve system availability and performance, and reduce the number of completions. It is also possible to improve the quality of projects and timeframes, to provide guidance and statistical data services in the required form.

Table 1. Benefits of ITIL

For customers (users)	For IT organizations
Providing of customer-oriented IT services, customer service quality agreement helps to improve relationship.	Identifying and optimization of a clear structure of IT department, focusing on corporate objectives.
Description of services in the customer language including required details.	Goal-orientation of the organization.
Improving quality control and services cost.	Providing the best conditions for effective outsourcing of IT services.
Improving the relationship between customers and IT organization.	Change of corporate culture in service as the main task of the IT department and supporting the implementation of a quality control system based on ISO-9000 standards.

A possible cause of failure of ITIL can be lack of certain processes [8]. ITIL defines the objectives and activities, input and output parameters of each process in the IT organization. However, these activities differ in each organization. Benefits of ITIL are presented in Table 1. Possible problems with ITIL may be as follows. Transition to the principles of ITIL can take a long time; require significant effort and changes in corporate culture. If improving processes structure is the goal, quality of services can be reduced. Improvement can not be achieved with a lack of understanding of goals and criteria for evaluating the effectiveness of processes. Lack of visible improvements in services and reducing costs. Necessity of involving management and employees of all organizational levels for successful implementation. Necessity of additional resources and staff.

Priority setting is an important aspect of business process management (BPM). Processes most critical to business have to be optimized primary [9]. The main idea of BPM is increasing operational efficiency through continuous improvement of processes.

To achieve this goal, it's necessary to get integrated tools and techniques that support a single-cycle management of the following tasks: processes design, their implementation and operational management processes, performance monitoring, analysis of statistics to identify reasons of the lack of efficiency of processes, setting goals to improve processes.

Methods of automation, execution and control are used for business processes automating in Business Process Automation System or BPM-, BPMS-Systems (Business Process Management System).

BPMS tasks include: interpretation and execution of a process model, conducting the process using its business rules; providing for users of a system a single point to access tasks and forms, access control and permission control, data verification on the assignments forms; managing the interaction of various enterprise information systems within business processes, control of information transfer between systems and process entities; providing an administrative interface for monitoring, online analysis and process monitoring, correcting the "problematic" process; providing historical information about the source of business processes and other process data for its further use in analytics.

An important trend in BPMS development is also a service-oriented approach. SOA (service-oriented architecture) as an approach allows us to consider all components of IT-architecture as a supplier (customer) service. SOA benefits include: reuse of the developed services and associated services directory providing the IT-environment changes effects prediction. SOA can be used together with the methodology of BPM, considering the service as a component of business processes and business processes as a whole, creating the concept of business services themselves.

3   Key Factors in Selecting Project for Business Process Management

Gartner analysts defined several key factors that should be taken into account when selecting the project for business process management (BPM) [4]. The main factor is that project should be created accordingly to the company goals. The process of IT modernization at the enterprise can be represented by the following figure (Fig. 1).

Data processing was carried out by a questionnaire research of businesses (users of IS), and IT companies in order to analyze the process of selecting and evaluating the effectiveness of IS in specific business environment. We can distinguish the following stages of the study [10]:

1. Editing and encoding information to formalize the information obtained during the study.

2. The choice of variables to study based on information gathered through questionnaires.

3. Statistical analysis was designed to identify patterns and relationships between the variables involved in the study.

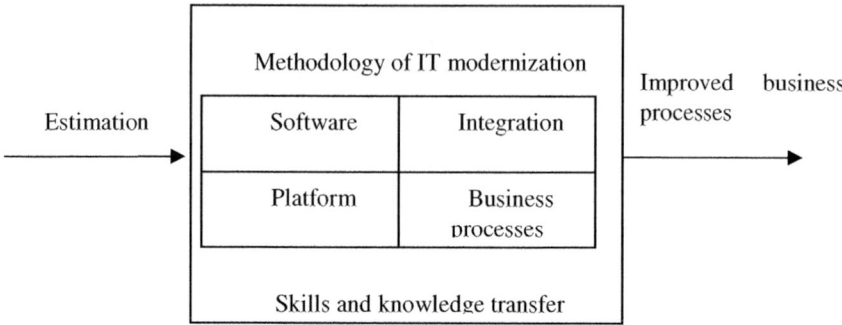

Fig. 1. The process of IT modernization at the enterprise

The survey was conducted on the basis of expert evaluation. For statistical analysis of data obtained during an interview, it was assumed that respondents are also experts in the field of knowledge. To estimate random errors or deviations in the analysis results there were used special methods for estimating the error of calculation. The questionnaire, which is a methodology of assessing the IS effectiveness at the enterprise is presented below.

Questionnaire research participant

"The practice of using IT in enterprises and companies of Ukraine"

Company (optional) _____
Thank you for participating in our study. Your answers are strictly confidential and will only be used for statistical analysis.

I. Company Overview

This part of the Questionnaire describes the sphere of a company's activity, its average turnover, the total approximate number of employees and IT staff. Such information is required to obtain a complete picture of the company's scope and area of business activity.

II. Relationship between Business and IT

The second part describes the important aspects in relationship between Business and IT including three subitems indicating the role of IT department in organization's structure, certain priorities of interaction with business and approaches used when making decisions on major business projects. Fragment of the second part is shown below.

2.1. What role does the IT department have in your organization? (Check one option).

	Technology division, responsible mainly for the work of specific devices and software
	Division responsible for providing support and IT services, ie specific business tasks
	Division actively offers solutions to problems business units and providing solutions to these problems in the form of IT services
	Direct part of the added value chains (directly create value for customers in the design, manufacture, sales, etc.)
	Hard to answer

2.2. What issues currently have the highest priorities in interaction with business? Check all answers.

IT costs:	
	development costs reducing while maintaining the number of IT projects
	operating costs reducing while maintaining the overall level of IT services
	reduction of IT staff
	reduction of labor costs without reducing the (significant) number of IT staff
	reduction of IT assets, who remain on the balance sheet and are available to IT departments
	severe reduction in IT costs in most of these areas
	on the contrary, increased spending on IT
Interaction with management and IT target setting:	
	strengthening the administrative-command form of management, the transition to direct orders management, more rigorous obedience IT management of business units
	transition into the language of management / financial accounting, improving control and transparency of costs of IT services (eg, due to substantiate claims of complete and detailed cost estimates, audit work and costs)
	increase the flexibility of the interface "Business - IT" (for example, the introduction of the catalog of IT services, including assessment of service costs, improving methods of assessing the contribution of IT in business)
	increase the degree of independence the IT department, the adoption of more authority and responsibility, increasing the confidence of the business (for example, the transition to self-financing)
Key parameters of the IT productivity:	
	increase flexibility in the use of IT resources (for example, by outsourcing, shared computing resources, hiring on a temporary contract basis, etc.)
	reducing the time to respond to business needs
	improving the quality of IT services
	requirements for the IT productivity parameters do not change
	requirements for the IT operation parameters reduced
	other (please specify) _____

2.3. Which methods / approaches you lean when making decisions on major business projects using IT. Check all suitable answers.

	assessment is not conducted, the decision is made on the inner opaque criteria
	project support from the business customer is considered a sufficient criterion value
	evaluated the qualitative results of the project (providing pass-through account, a common database, etc.)
	estimated quantitative impact of the project on indicators (KPI) of business processes (for example, shortening estimates, reducing the frequency of errors, etc.)
	estimated financial payback of the project
	carried out a qualitative assessment of risks of the project (list of risks, risk management scenarios)
	conducted a quantitative risk assessment of the project (calculated as the probability of success of the project)
If during the evaluation of the project model is used, please specify the type(s):	
	best practice in IT operational efficiency ensuring (ITIL, etc.)
	methods for determining the full cost of the acquisition, use and disposal systems (TCO, etc.)
	techniques associated with the interaction of IT and business (CoBIT, ValIT, etc.)
	methodology for quantifying the business as a whole (MOS, KGI / KPI, BSC, etc.)
	methodology for assessing economic performance of IT projects (EVA, ROI, TVO, AIE, REJ, etc.)
	other models and methodologies (please specify) _____

III.   IT Budget and Finance

The third part of the Questionnaire describes the finance sphere, including a company's IT budget in percents to the total company's turnover and the distribution of IT budgets among different types of costs, divided by a work type. They are as following: automation projects in non automated areas (implementation of new business applications), IS development projects (development and modification of existing applications), deployment of new IT infrastructure, Support and maintenance of existing IT systems, development of IT management (standardization and methodological support) and other.

IV.   Tasks and Areas of Work in the IT

The fourth part of the Questionnaire describes the importance of the following areas of business processes automation: Marketing (market analysis, planning and forecasting, pricing); Development of products and services (research, development, preparation of production processes); Production of products and services (planning and production accounting, change management, quality, equipment maintenance); Supply, distribution, delivery (supply chain management and inventory management, supply and delivery of products and services); Sales (order management and payments, customer relationship, service support and claims or refunds); Finance (tax and management accounting, cash flows, settlement, cost, budgeting, reporting); Staff (personnel records, salaries, hiring, compensation and training); Planning (strategic and operational planning, program management, monitoring plans); Improving the performance of the company and its business processes (business process analysis, management of KPI).

There are also studied certain projects that have already been initiated or planned to initiate in the next six months. The automation functions are divided into areas and sub-areas of business processes as shown before.

The information gathered can be used to implement the company's tactics of efficient development.

4   The Use of Multiple Regression Models to Survey Data

The aim of studying the set of companies, ranked by the level of IT-departments competitiveness is to evaluate the nature of the influence of a factor in creating and maintaining a strong IT-industry.

For determining the appropriate weight categories and indicators were used survey results of businesses (see Table 2).

Table 2. Source data. Rating of subsystems enterprises by importance in terms businesses.

Area of business processes / Enterprise	Key business activity	Performance Indicator (y)	Marketing (x_1)	Development (x_1)	Production (x_2)	Supply (x_3)	Sales (x_3)	Finance (x_4)	Planning (x_5)	Staff (x_6)
Ltd. "KMT-Enegry"	Energy service	1	2	2	4	4	5	3	1	1
Ltd. Agrocom-Invest	Agriculture	1	1	1	5	3	1	5	3	3
Beauty salon "Tandem"	Health care service	3	0	0	0	3	2	5	5	2
Ltd. "Modern food technology "Citron"	Food industries	4	3	5	4	3	5	5	4	2
Kharkov Industrial Meat Complex	Food industries	4	3	1	5	2	2	5	4	1
Scientific Production Corporation FED	Engineering company	3	3	2	5	4	2	4	4	1

Table 2. (*continued*)

Kharkov State Aircraft Manufacturing Company	Aircraft manufacturing	4	5	4	3	3	2	3	4	2
Nitrolabs Internet Technologies Laboratory	Software development	2	2	1	3	3	5	5	4	5
State Scientific Institution "Institute for Single Crystals" of National Academy of Sciences of Ukraine	Physics, chemistry, biology and medicine interdisciplinary scientific researches	4	2	1	3	2	4	5	4	5

To study the subrating companies in order of importance in terms of business, we constructed a multiple regression model. The solution to this task was implemented using a software package Statistica 6.0.

Let's find the relationship between data, determining the impact of factors on the companies IT subsystems rating. There was used the correlation coefficient, which shows the strengthness of the relationship between variables.

Despite the fact that one of the conditions of a regression model is the assumption of linear independence of the factors, this condition for economic calculations does not always happen. To reduce the degree of dependence between factors there was conducted their staged exclusion from the model, as well as replacement of the factors similar in economic significance.

Multifactor linear regression model is described by the equation:

$$y = a_0 + a_1 * x_1 + a_2 * x_2 + \ldots + a_n * x_n + e \tag{1}$$

The subsystems rating was used as the result factor. It is also necessary to verify the absence of significant associations between other factors that are considered independent ($x_1\ x_2\ x_3 \ldots x_n$).

Thus, the first model (column Beta), which takes into account all factors and, without a constant term in the general form:

$$y = a_1 x_1 + a_2 x_2 + a_3 x_3 + a_4 x_4 + a_5 x_5 + a_6 x_6 \tag{2}$$

The model was obtained using Statistica 6.0. The calculated coefficients:

$$y = 0{,}71 x_1 - 0{,}34 x_2 - 0{,}48 x_3 - 0{,}02 x_4 + 0{,}45 x_5 - 0{,}41 x_6 \tag{3}$$

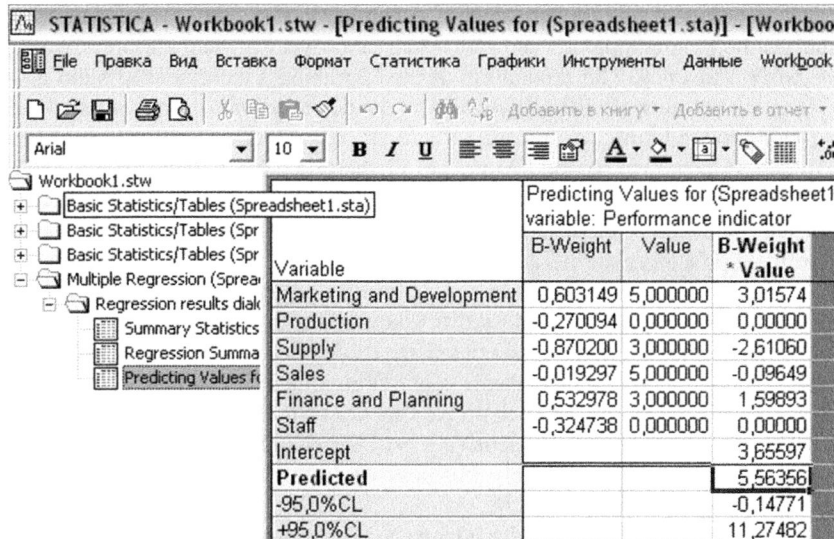

Fig. 2. Prediction model result

The second model (column B), taking into account all coefficients (a) model, in a general form:

$$y = a_0 + a_1x_1 + a_2x_2 + a_3x_3 + a_4x_4 + a_5x_5 + a_6x_6 \tag{4}$$

Using the coefficients:

$$y = 3,66 + 0,60x_1 - 0,27x_2 - 0,87x_3 - 0,02x_4 + 0,53x_5 - 0,32x_6 \tag{5}$$

We construct a prediction model obtained by using (2), as shown on fig.2. Predictive value is contained in the Predicted, box B-Weight - a weight, Value - x, + -95,0% CL - confidence level where the values are with 95% probability.

Thus, we can conclude that the model shows a good relationship with the variable (the IT efficiency indicator) on the independent variables x1, x2, x3, x4, x5, x6 (marketing and development, production, supply and sales, finance and planning, staff) and can be used for prediction.

The resulted model shows that company managers mostly prefer using the subsystem of marketing and development, considering it carries the greatest impact on business efficiency. So, using the model described, we may predict certain level of the business efficiency of an Information System according to the expert opinion of managers questioned.

5 Conclusion

Information technologies become an increasingly important factor for economic and social development in all countries. In most developed countries advanced IT-industry creates conditions for the continued strengthening of the economies of these countries, helping companies and individual employees to increase productivity and raise work efficiency. The orientation of the Ukrainian government to build an innovative economy must significantly promote the development of innovative climate in the country in the future.

To ensure long-term competitiveness of IT sector the key parameters are: the quality of local technological infrastructure, the availability and level of training of IT specialists, innovative environment, regulation, and business conditions, as well as government policies in the field of technology.

We studied a range of Ukrainian companies to find out the type of dependence between level of company's business efficiency and use of information technologies. During the investigation there was developed a detailed questionnaire, which results allowed to build a model of multiple regression, aimed to examine certain relations between Business and IT. The constructed model of multiple regression allows us to trace the dependence of the competitiveness of the IT industry on factors related to its development and support, namely business environment, IT infrastructure, human capital, support for IT development.

According to the criteria of the adequacy and accuracy of the model considered adequate and accurate enough for analysis. Using this model there were also calculated prediction values of the dependent variable representing the level of the effectiveness expected in the overall ranking.

The findings of the study, helped to create the necessary conditions for strengthening the IT sectors of their economies and thereby take advantage of social benefits and advantages provided by a high level of IT development.

The final decision for choosing a particular Information System can be obtained, for example, by analyzing the average estimates, which domain experts from the company elected to put the IS for compliance with these criteria [11, 12, 13, 14]. Thus, the application of the proposed model assessing the effectiveness of IS can significantly improve the financial results of the implementation project.

References

1. Sizov, A.: Designing an Enterprise Architecture: selecting the IT tools. CNews-analyst (August 23, 2010),
   http://www.cnews.ru/reviews/index.shtml?2010/08/23/-406142_1
2. Zachman, J.A.: A Framework for Information Systems Architecture. IBM Systems Journal 26(3), IBM Publication G321-5298,
   http://www.research.ibm.com/journal/50th/applications/zachman.html
3. Institute For Enterprise Architecture Developments, http://www.enterprise-architecture.info/ifead%20about.htm
4. Zeltser, A.: Gartner: seven conditions of BPM project success. It-World 07(147) (April 2010),
   http://it-world.ru/upload/iblock/300/92025.pdf
5. Kadushin, A., Mikhailova, N.: Methodology of evaluation the economic efficiency of IT projects. IF-Consult (July 07, 2003),
   http://www.pmprofy.ru/content/rus/83/833-article.asp

6. Strassmann, P.A.: Why ROI ratios are now crucial to IT investment? Butler Group Preview (September 2002)
7. Tselykh, A.: Ltd. "KORUS Consulting". Evaluation of IT-projects. A balanced approach, http://quality.eup.ru/MATERIALY5/oe-it.htm
8. Reznichenko, G.: The IT management - the art of achieving business goals, http://www.pronet.kiev.ua/press/articles/BMC-ITandBusiness.pdf
9. Regional features of process management. Open Systems (April 13, 2011), http://www.osp.ru/news/2011/0413/13006816/
10. Falkenberg, G.: Enterprise Application Development and Maintenance. Natural and Adabas Products (November 2007), http://www.mops1.com/softwareag/bid/pdf/Guido%20-Falkenberg.pdf
11. Negomedzyanova, E.: Forming the model of estimation the economic efficiency of the generating company. Journal of Scientific PhD and Doctoral Publications (6) (2007), http://www.jurnal.org/articles/2007/ekon43.html
12. Rampersad, H.K.: Total Performance Scorecard: Redefining Management to Achieve Performance with Integrity, p. 336. Butterworth-Heinemann, Oxford (2003) ISBN: 978-0-7506-7714-1
13. Walsh, C.: Key Management Ratios: How to Analyze, Compare and Control the Fig-ures that Drive Company Value (Financial Times Management Masterclass Series), International edition, p. 347. Financial Times/Prentice Hall, ISBN-13: 978-0273621973
14. Walsh, C.: Management Ratios: The Clearest Guide to the Critical Numbers That Drive Your Business, p. 416. Companion Group (2006) ISBN 966-96692-0-0, 0-273-70731-0

Part III

Software Development

A Method to Discover Trend Reversal Patterns Using Behavioral Data

Jerzy Korczak and Aleksander Fafuła

Wrocław University of Economics, Poland
{jerzy.korczak,aleksander.fafula}@ue.wroc.pl

Abstract. Cognitive biases often influence decision processes related to investment on stock markets. Mainly, this concerns complex problems with perception and understanding of surrounding financial and economic reality. This research was aimed at detecting cognitive biases in the data-driven manner. A few basic cognitive biases were examined: the Gambler's Fallacy, and the Hot Hand and Cold Hand effects. Detecting and modeling sequences leading to particular cognitive bias can significantly improve the trader's strategy.

This paper presents a concept of a platform which can detect specific user behaviors. These are derived from the observation of technical analysis indicators, as well as the trader's own built-in indicators. Along with the standard functionalities of a stock market simulator, a few methods of data mining were applied: inductive decision trees, sequential association rules, clustering and visual exploration.

Keywords: cognitive biases, trading support software, patterns in behavior, trend reversal.

1 Introduction

Is it possible to alert traders, that they are going to make a mistake tomorrow? The science of Investment Psychology points out various cognitive biases. These have been widely tested for many decades. Due to the nature of the matter some models of behavior are better known than others. In this study, a few of them have been chosen as they appear to be more common, namely:

1. The Gambler's Fallacy – the belief in negative autocorrelation - the tendency to think that future probabilities are altered by past events, when in reality they are unchanged [1]. This results from an erroneous conceptualization of the law of large numbers. For example, if an abundance of heads comes up in tossing a coin, observers may be heard to assert that tails are due; that tails are more likely to come up than another head [2]. On the stock market this can be accomplished by holding losing stocks for too long and selling winning stocks too quickly [3]. Gamblers holding stocks for too long expect a reversal.
2. The Hot Hand Effect – the belief in positive autocorrelation – It was first demonstrated by Gilovich, Vallone and Tversky [4], that people believe in the hot hand in basketball shooting (striking more makes a player hot). On the stock market it can be accomplished by selling "losers" and buying "winners". DeBondt and Thaler [5]

S. Wrycza (Ed.): SIGSAND/PLAIS 2011, LNBIP 93, pp. 81–91, 2011.

showed that customers who rely on past performance are overly optimistic about past winners and overly pessimistic about past losers.

These two biases are not simply opposites. In the Gambler's Fallacy the coin is due, in the Hot Hand the person is hot. This difference is also similar for the stock market, for example the Gambler's Fallacy can reflect the price, and Hot Hand can apply for companies.

A full list and extensive descriptions of experiments on cognitive biases detection can be found in the References, below. However, cognitive science has not been introduced to data-mining on a scale provided by computing power and network availability. It is unclear and thus needs further investigation if a data-driven approach is able to give additional perspectives for the detection of cognitive biases.

This paper presents a brand new concept of a system which can help improve traders by reducing the proportion of cognitive biases. The simulator itself applies the data-driven approach to the detection of cognitive biases. To make results clearer, the study focuses on newbie investors. The main question of this project is the following: is it possible to detect cognitive biases automatically?

We will demonstrate that it is also possible to use this knowledge as an early signal of trend reversal. In talking about trend reversals, we don't focus on any type of reversal in particular. This can be either intra-day or inter-day situation. This research has also different approach than existing indicators like ROC, MACD, Moving Averages or Williams%R. Therefore, knowing the sequence of patterns should produce indicator that is neither leading nor lagging. The aim of this research is to track patterns in data directing to the given situation. From now on, we focus on the Gambler's Fallacy effect occurring when buying stocks. We assume that patterns introducing Gambler's Fallacy are confirmed, when there is a trend reversal in a couple of days following. This should be caused by investors falling into the trap of faulty judgment. A belief in a sudden price increase and extensive buying of stocks should (in theory) lead to an increase of volume and price itself.

Moreover this paper describes briefly six stages to build and deploy the trading platform which can track users' behaviors and inform them about the cognitive biases. The technology is clearly defined, based on the authors' experience. This early methodology should be treated as the first debut in this area of research. At the beginning, the paper describes technical issues, along with the main algorithm. Subsequently, some conclusions are drawn and affiliations with the virgin run of the prototype of the platform are made.

2   Description of Experiments and Applied Methods

The research were done on a platform designed to collect behavioral data from traders. The most important elements of the system are presented in the diagram (Figure 1).

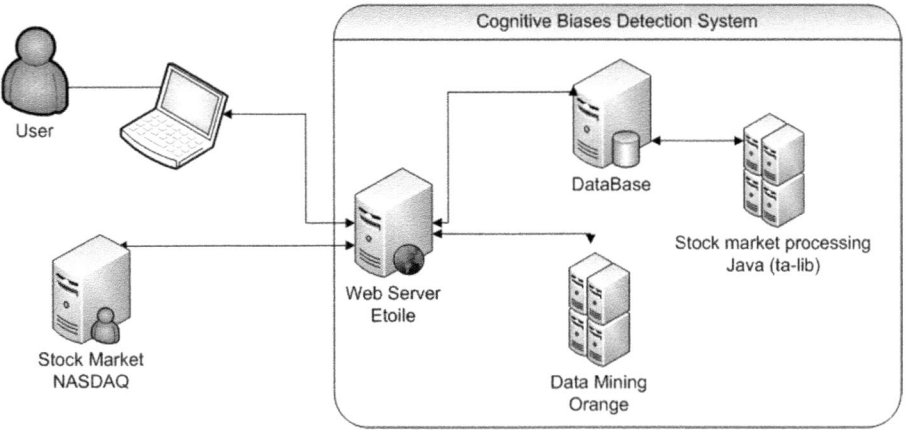

Fig. 1. Elements of the platform with their interactions

The diagram of the platform and basic data flows do not reflect the detailed structure, but provide the reader with the broad view of the architecture. Elements included and used by the platform are: user interface (served by web server with the web application Étoile including quizzes and stock market simulator's controls); layers of the data (i.e. cognitive patterns, users' behavioral data, stock quotes stored in database), data mining facilities, and stock market computing software. The following section of this paper describes these in detail, proposing a very early methodology for creating the platform to detect cognitive biases using a data-driven approach.

The experiments were conducted in a form of investment game. It has been run three times during the lab classes on the Wrocław University of Economics, Poland. Students were also able to continue the task from their own computers, because the system was operating 24/7. Every game took about one month, and has been conducted in years 2009, 2010 and 2011. The total amount of collected buy and sell signals is about 2000. The real number of usable entries differs, since various filtering criteria can be applied to clean up the data. In every experiment students were put into a stock broker role and operated on NASDAQ (from now on we refer to them as: users, investors or brokers). Finally, the game has been completed by about a hundred users. Special stock market simulator has been written and constantly improved to provide specific functionalities.

The experiment were carried out in the following stages:

1. Platform development.
2. Processing of quizzes.
3. Stock data processing.
4. Data mining process using Orange.
5. Interpretation of the results.
6. Testing newly found rules.

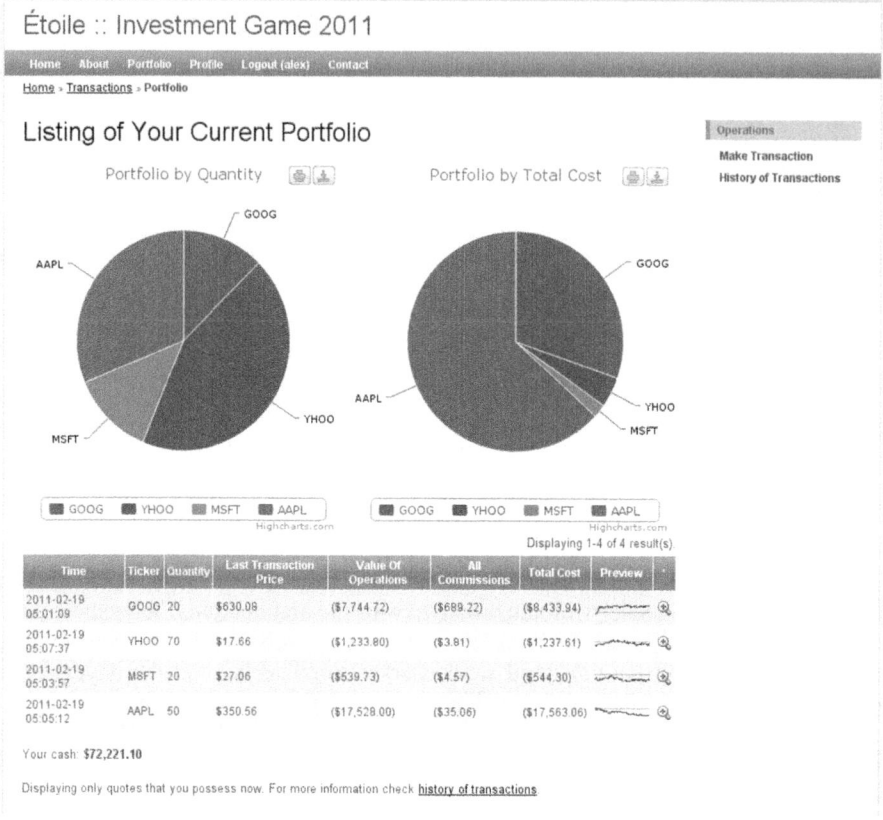

Fig. 2. User's portfolio in Étoile

Stage 1

As a platform a YII framework was chosen because it utilizes the MVC architecture. The investment game is intended for beginning traders, because they more often have less experience and are thus more susceptible to cognitive biases.

The game itself looks like a trading simulator. It was designed to be transparent for the gamer. The participants are not being warned that this test is about psychology. Questions are displayed randomly and they are not obliged to answer them at all.

After successful logon, users see their portfolio. The interface has been simplified to improve usability as well as learnability. There are also charts displaying portfolio by quantity and total cost which is show in fig. 2.

Students who would like to make a transaction are redirected to the form, where it is possible to buy or sell stocks. After successful completion of this stage (validation against entered quantities and ticker existence) the quiz is presented to the user, as shown in Fig. 3.

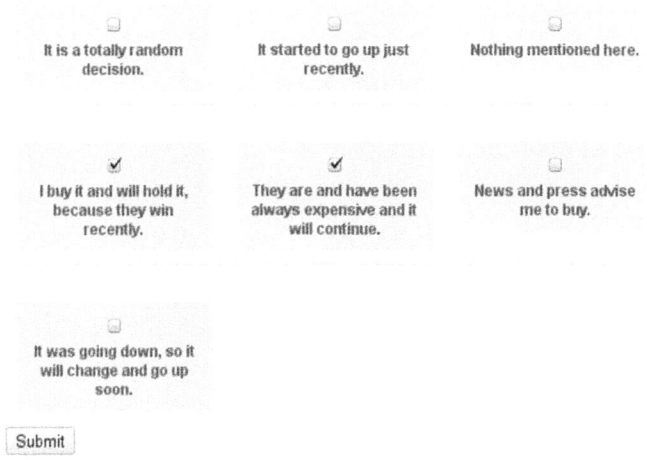

These questions are not a part of assingment, but additional game for students. Results will be shown at the end of game. Students providing answers will receive more feedback and improve their market skills.

Please choose sentences, describing your decision, that you feel you agree with:

Fig. 3. Example survey after "buy" decision

Questions are randomly selected from the database. They are categorized for buying or selling. They also fall into given cognitive problem theory.

Stage 2
The matrix of stock data and human behavior is created with the use of questionnaire. Questions are finally coded into four classes: "Gambler's Fallacy" (buy/sell) or "Hot Hand" (buy), "Cold Hand" (sell). For example statement: *"It is expensive, so will be cheap soon"* will be coded as "Gambler's Fallacy" (cf. Fig. 4).

```
case "It is expensive, so will be cheap soon.":
case "It is going too much up so will change.":
case "I see on the chart that the price rises, so it will change and fall.":
case "I am going to benefit from the just started rising trend.":
case "The trend is rising too long, so there is a risk of a change.":
case "I see on the chart that price goes down, so it will change and go up.":
case "This company is not very good, so they will improve.":
case "It was going down, so it will change and go up soon.":
case "The rising trend has started not long ago.":
case "It was going up, so it will go down soon.":
case "It started to go up just recently.":
  $gamblers = $gamblers + 1;
```

Fig. 4. Coding indicators for Gambler's Fallacy

At the filtering stage it is possible to remove answers that are random, not relevant, or false. For instance, a student may state that he/she is buying because the price is falling and rising, and this simply indicates that we should not include the answers in the calculation.

Stage 3

The data part is constructed from technical analysis (momentum, overlap volume, cycle, self-made etc). They are calculated using the daily data. Finally they are decoded from numerical values to their interpretation, as show in the figure below.

```
public function ionsma($fast, $slow)
{
  if ( $fast > $slow )
  {
    $interpretation = "sma go long";
  }

  if ( $fast < $slow ) {
    $interpretation = "sma go short";
  }

  if ( $fast == $slow ) {
    $interpretation = "sma crossing";
  }

    return $interpretation;
}
```

Fig. 5. Technical analysis coding

Stage 4

Data from the stock simulator are transferred to the Orange[1] tool. The elements of a sample export file are shown in Fig. 6 and 7. Note that tables presented in Fig. 6 and 7 do not contain all headers used. This is limited due to the format of this publication.

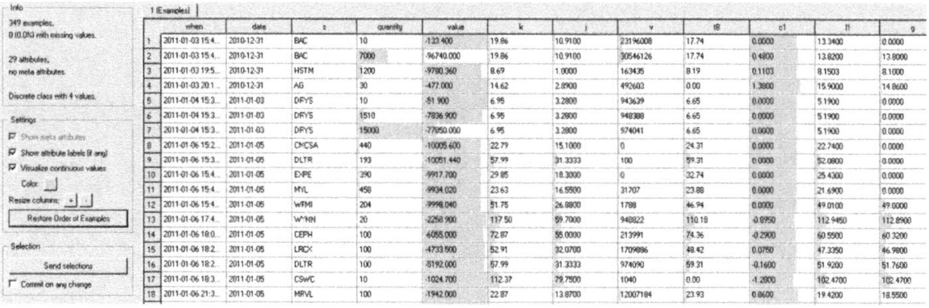

	when	date	s	quantity	value	k	j	v	t6	c1	f1	g
1	2011-01-03 15:4...	2010-12-31	BAC	10	-133.400	19.86	10.9100	23196008	17.74	0.0000	13.3400	0.0000
2	2011-01-03 15:4...	2010-12-31	BAC	7000	-96740.000	19.86	10.9100	30546126	17.74	0.4800	13.8200	13.8000
3	2011-01-03 19:5...	2010-12-31	HSTM	1200	-9780.360	8.69	1.0000	163435	8.19	0.1103	8.1503	8.1000
4	2011-01-03 20:1...	2010-12-31	AG	30	-477.000	14.62	2.8900	492603	0.00	1.3800	15.9000	14.8600
5	2011-01-04 15:3...	2011-01-03	DFYS	10	-51.900	6.95	3.2800	943639	6.65	0.0000	5.1900	0.0000
6	2011-01-04 15:3...	2011-01-03	DFYS	1510	-7836.900	6.95	3.2800	948388	6.65	0.0000	5.1900	0.0000
7	2011-01-04 15:3...	2011-01-03	DFYS	15000	-77850.000	6.95	3.2800	974041	6.65	0.0000	5.1900	0.0000
8	2011-01-06 15:2...	2011-01-05	CMCSA	440	-10005.600	22.79	15.1000	0	24.31	0.0000	22.7400	0.0000
9	2011-01-06 15:3...	2011-01-05	DLTR	193	-10051.440	57.99	31.3333	100	59.31	0.0000	52.0800	0.0000
10	2011-01-06 15:4...	2011-01-05	EXPE	390	-9917.700	29.85	18.3000	0	32.74	0.0000	25.4300	0.0000
11	2011-01-06 15:4...	2011-01-05	MYL	458	-9934.020	23.63	16.5500	31707	23.88	0.0000	21.6900	0.0000
12	2011-01-06 15:4...	2011-01-05	WFMI	204	-9998.040	51.75	26.8800	1788	46.94	0.0000	49.0100	49.0000
13	2011-01-06 17:4...	2011-01-05	WYNN	20	-2258.900	117.50	59.7000	948822	110.18	-0.8950	112.9450	112.8900
14	2011-01-06 18:0...	2011-01-05	CEPH	100	-6055.000	72.87	55.0000	213991	74.36	-0.2900	60.5500	60.3200
15	2011-01-06 18:2...	2011-01-05	LRCX	100	-4733.500	52.91	32.0700	1709886	48.42	0.0750	47.3350	46.9800
16	2011-01-06 18:2...	2011-01-05	DLTR	100	-5192.000	57.99	31.3333	974090	59.31	-0.1600	51.9200	51.7600
17	2011-01-06 18:3...	2011-01-05	CSWC	10	-1024.700	112.37	79.7500	1040	0.00	-1.2800	102.4700	102.4700
18	2011-01-06 21:3...	2011-01-05	MRVL	100	-1942.000	22.87	13.8700	12007184	23.93	0.8600	19.4200	18.5500

Fig. 6. Left side of the export preview table

[1] `http://orange.biolab.si/`

macdhist	macdhistdr	mfi	wilt	sma	sar	bbands	obv	atr	familiar	possible-fallacies
macd histogram p...	macd hist falling	mfi normal (do not...	w%r overbought (...	sma go short	sar go long	bbands middle to...	obv rising	atr normal	not known	Hot Hand
macd histogram p...	macd hist falling	mfi normal (do not...	w%r overbought (...	sma go short	sar go long	bbands middle to...	obv rising	atr normal	not known	no fallacies
macd histogram p...	macd hist rising	mfi overbought (s...	w%r normal	sma go long	sar go long	bbands above up...	obv rising	atr elevated	not known	Hot Hand
macd histogram p...	macd hist rising	mfi normal (do not...	w%r overbought (...	sma go long	sar go long	bbands above up...	obv rising	atr normal	not known	collision of fallacies
macd histogram n...	macd hist falling	mfi oversold (buy)	w%r oversold (buy)	sma go long	sar go short	bbands below lo...	obv falling	atr normal	not known	no fallacies
macd histogram n...	macd hist falling	mfi oversold (buy)	w%r oversold (buy)	sma go long	sar go short	bbands below lo...	obv falling	atr normal	not known	no fallacies
macd histogram n...	macd hist rising	mfi normal (do not...	w%r overbought (...	sma go long	sar go long	bbands middle to...	obv rising	atr normal	not known	Gambler's Fallacy
macd histogram n...	macd hist falling	mfi normal (do not...	w%r oversold (buy)	sma go long	sar go short	bbands below lo...	obv falling	atr elevated	not known	Gambler's Fallacy
macd histogram n...	macd hist falling	mfi normal (do not...	w%r normal	sma go long	sar go long	bbands middle to...	obv falling	atr elevated	not known	Hot Hand
macd histogram p...	macd hist rising	mfi normal (do not...	w%r overbought (...	sma go long	sar go long	bbands middle to...	obv falling	atr normal	not known	Gambler's Fallacy
macd histogram n...	macd hist falling	mfi normal (do not...	w%r oversold (buy)	sma go long	sar go short	bbands middle to...	obv falling	atr normal	not known	Hot Hand
macd histogram p...	macd hist falling	mfi overbought (s...	w%r overbought (...	sma go long	sar go long	bbands above up...	obv rising	atr normal	not known	no fallacies
macd histogram n...	macd hist falling	mfi normal (do not...	w%r oversold (buy)	sma go long	sar go short	bbands below lo...	obv falling	atr normal	not known	no fallacies
macd histogram n...	macd hist falling	mfi normal (do not...	w%r oversold (buy)	sma go long	sar go short	bbands below lo...	obv falling	atr elevated	not known	no fallacies
macd histogram n...	macd hist falling	mfi normal (do not...	w%r normal	sma go long	sar go short	bbands middle to...	obv falling	atr normal	not known	no fallacies
macd histogram n...	macd hist rising	mfi normal (do not...	w%r normal	sma go long	sar go short	bbands middle to...	obv falling	atr normal	not known	Hot Hand

Fig. 7. Right side of the export preview table

The export file is processed off-line, to find decision tree rules that connect the market situation with the classes and their relevant cognitive biases. The description of data-mining algorithms is beyond the scope of this document, but the brief preview of the architecture is represented by Fig. 8. The icons are the part of visual programming process called widgets. Each of them performs some task, for example: select data, build a classification tree, or display calibration plot. Connections between widgets represent data flows – communication channels. Sets of widgets and channels are known as schemas. The biggest advantages of using schemas are that they are: reusable, easy to understand and fine-tune.

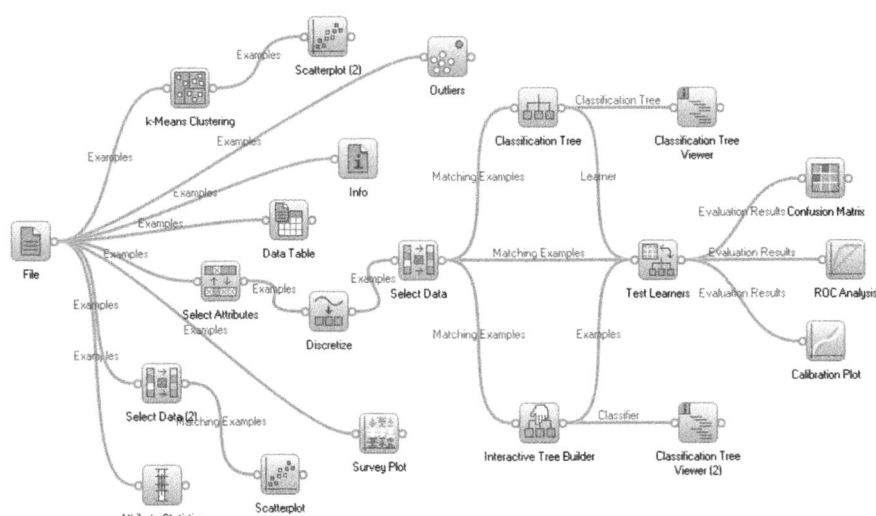

Fig. 8. Overview of data mining process

Stage 5

The interpretation of the results is made by linking rules found at the Stage 3 with the knowledge provided by the psychology. It is clear that some interpretations might be more accurate and better reflect the market reality than the others.

For instance (ceteris paribus) a rule indicating that the Gambler's Fallacy (in the situation of buying stocks) should not be associated with the high price. Alternatively, the Hot Hand effect should not be detected when the price is very low. Furthermore, the Gambler's Fallacy (when buying) should not be indicated by Bollinger Bands suggesting that the market is overbought.

As a matter of success, the experimental results did not produce a high rate of false positives. This is also the first proof of validity of the results (as shown on Fig. 9 to 11).

Stage 6

The implementation of the rules found during the Stage 4 is performed manually. Warnings are displayed to the user while trading in the game in the scope of a given rule. By way of brief illustration – one of rules that was found (simplified):

```
If (     ( Bollinger Bands = Middle to Lower ) AND
   ( Williams%R = Normal ) AND
   ( MACD Histogram Direction = Falling ) )
Then
   ( Possible fallacies = Gambler's Fallacy )
```

Rules are pre-evaluated by special database module. If they pass this stage they are coded in MQL4 language for further investigation. Then, they are tested with the real (tick) data, using the maximum accuracy possible.

3   Discussion of Results

The prototype *Étoile* has been already tested three times. This was achieved during the games organized for students of the Wrocław University of Economics, Poland. The sample collected in previous trials with the use of *Étoile* shows some encouraging information, notably the first examples of efficient behavioral rules [6]. Thus it is clear that wider experiments should be carried out. Results show that the decisions of the users, supported and related to answers to the simple survey, can indeed create patterns. Discovering rules for cognitive biases is very challenging and creates a broad area of research, especially in the domain of interpretation of the results.

In Fig. 9 to 11 some early results are presented that have been achieved using behavioral data hybrid indicators. This algorithm finds occurrences of patterns leading to strong beliefs in an event of a trend reversal. So the strategy is to buy before this known pattern ends, i.e. a large number of investors would actually buy and therefore provoke a price rise. In the examples provided, the behavioral indicators compute only starting point (buy signal). The exit point is calculated using various and arbitrary technical analysis indicators.

Fig. 9. The chart of Microsoft Corporation (MSFT) with behavioral transaction

Fig. 10. The chart of Intel Corporation (INTC) with two buy signals (weak and strong)

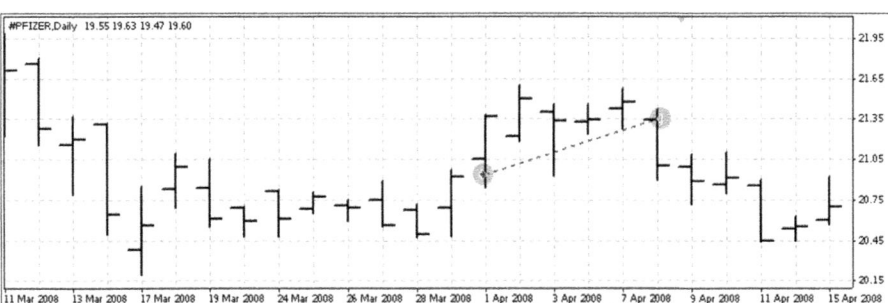

Fig. 11. The chart of Pfizer Inc. (PFE) zoomed

As shown on the charts, the precedomg price movements may be described as: fluctuating low with falling tendency, formerly much higher. Despite the fact of the real gain, or calculating the exit point, results show that the ratio of price rise is very high. This is presented by the Table 1. The average trading time was only 7-12 days each. Due to diversity of investing means the performance of this indicator was measured by the ratio:

(price raise after the buy signal = success)

/

(price fall after the buy signal more than 6% = fail)

Table 1. Examples of ratio performance and profit factor

	MSFT	INTC	PFE
Total trades	3	5	5
Long positions won	3 (100%)	3 (60%)	4 (75%)
Long positions loss	0	2 (40%)	1 (25%)
Profit factor	n/a	5.87	6.97

The results have been calculated since the beginning of 2007 until July 2011 using tick data. Despite the fact that this indicator provides rare buy signals it can be used as a market scanner. Finding more companies in detectable situations enables investors to use this indicator almost constantly.

For more information regarding current signals visit http://bdm.ue.wroc.pl/. In addition, a brief presentation with screen shots is available at: http://citi.ue.wroc.pl/etoile/p/.

4   Conclusions

After describing the proposition of a method to find trend reversal patterns with the experimental software, some basic data exploration approaches have been introduced. Despite the fact, that this is the first attempt of building the new indicator, the results are promising. The behavioral rules found, after fine tuning, are successfully becoming efficient hybrid indicators. Perhaps this will stimulate future research on creating new behavioral-based indicators.

Now, after three experimental market sessions on NASDAQ, some new features are introduced to our software: "live advisor" for investors - suggesting a risk of a particular fallacy occurring; "explain" - displaying the rules found, with explanation; and finally - the ability to extend the trading platform for a larger group of scientists and individuals willing to take a part in research and educational simulation games.

Finally, it is necessary to discuss this experiment with the wider audience, involving especially: psychologists, financial specialists and traders. This would enable the system to be extended and improved.

References

1. Tversky, A., Kahneman, D.: Belief in the Law of Small Numbers. Psychological Bulletin, 105–110 (1974)
2. O'Neill, B., Puza, B.D.: Dice Have No Memories but I Do: A Defence of the Reverse Gambler's Belief (2004); reprinted in abridged form as: O'Neill, B., Puza, B.D.: In defence of the reverse gambler's belief. The Mathematical Scientist 30(1), 13–16 (2004)
3. Odean, T.: Are Investors Reluctant to Realize Their Losses? Journal of Finance 53(5), 1775–1798 (1998)
4. Gilovich, T., Vallone, R., Tversky, A.: The hot hand in Basketball: On the Misperception of Random Sequences. Cognitive Psychology (17), 295–314 (1985)
5. DeBondt, W., Thaler, R.: Does the Stock Market Overreact? Journal of Finance 40, 793–805 (1985)
6. Fafuła, A.: A prototype of platform for data-driven approach to detection of cognitive biases. In: Korczak, J. (ed.) Data Mining and Business Intelligence. Research papers of Wrocław University of Economics No. 104, Business Informatics, vol. 16, pp. 71–78. UE, Wrocław (2010)

Toward a Theory of Debiasing Software Development

Paul Ralph

Lancaster University, Lancaster, UK
paul@paulralph.name

Abstract. Despite increasingly sophisticated programming languages, software developer training, testing tools, integrated development environments and project management techniques, software project failure, abandonment and overrun rates remain high. One way to address this is to focus on common systematic errors made by software project participants. In many cases, such errors are manifestations of cognitive biases. Consequently this paper proposes a theory of the role of cognitive biases in software development project success. The proposed theory posits that such errors are mutual properties of people and tasks; they may therefore be avoided by modifying the person-task system using specific sociotechnical interventions. The theory is illustrated using the case of planning poker, a task estimation technique designed to overcome anchoring bias.

Keywords: Design Science, Software Engineering, Theory Development, Cognitive Bias, Debiasing, Heuristics, Illusions.

1 Introduction

Software development and maintenance constitute substantial economic activity – the 500 largest software companies employed 3,562,407 and accrued revenues of US$491 billion in 2010 [1] and total global information technology spending was US$3.2 trillion in 2009 [2]. The magnitude of this spending makes estimates of failure rates far more alarming. Estimates of completely abandoned software development projects vary between 10% and 44% while between 16% and 52.7% experience "major truncation or simplification … prior to full implementation" [3, p. 17, 19]. More recently, the success rate has been estimated at 32% with 44% "challenged" [4]. A meta-analysis of estimation accuracy studies found an average effort overrun of 30-40% [5]. If civil-engineering projects had similar success rates, abandoned buildings and half-built bridges would litter our cities.

This raises many questions. What causes software project success, abandonment and failure? How should success and failure be measured? How can software project outcomes be improved? Which causes of failure are reparable and by whom? For the purposes of this paper, design (verb) refers to the act of creating a specification of an object intended to accomplish goals in a particular environment using a set of primitive components, subject to constraints where a specification may be a plan or the object itself (adapted from [6]). A software design project then "is a temporal trajectory of a work system toward one or more goals" [6, p. 116] where at least one goal entails creating software.

S. Wrycza (Ed.): SIGSAND/PLAIS 2011, LNBIP 93, pp. 92–105, 2011.
© Springer-Verlag Berlin Heidelberg 2011

Van de Ven [7] argues that completely understanding such complex phenomena requires multifarious theoretical perspectives. This paper proposes a cognitive psychology perspective – specifically, using the notion of cognitive biases to understanding *avoidable* errors and how to address them in software design projects. In summary, the purpose of the paper is as follows.

> **Purpose:** to propose a theoretical framework of errors in software development projects and their relationship to cognitive biases.

To this end, the paper is organized as follows. Section Two provides a brief review of existing research on antecedents of project success and failure. This is succeeded by an introduction to the basis for the proposed theory – cognitive biases (§3). The proposed theory is presented in Section Four and evaluated in Section Five.

2   Existing Perspectives on Critical Success Factors in Software Development

Both scientific and popular literature are rife with examples of high profile design failures. Connected Earth's new million-pound website was incompatible with nearly all modern web browsers [8]. The British supermarket Sainsbury's spent US$526 million on a supply chain management system that failed so dramatically that the company had to hire 3000 people to move stock manually [9]. A recent study of a computerized physician order entry (CPOE) system intended to reduce medication errors found that the system caused patients to get the wrong medicine in 22 different ways [10]. When Windows Vista was first released, consumers were so unhappy that "more than one in every three new PCs [was] downgraded from Windows Vista to the older Windows XP, either at the factory or by the buyer" [11, p. 1].

Many studies have listed diverse success, abandonment and risk factors for software projects. Table 1 provides three exemplars. While identifying these factors is useful to predict success and identify high-risk projects, many of the factors are beyond project participants' control. For example, changing requirements and unclear objectives may be unavoidable. The most striking example is perhaps "executive management support". Noting low executive support is important for gauging risk but it is not necessarily actionable. More generally, identifying factors associated with success and failure is not equivalent to demonstrating causal relationships.

The non-actionable nature of many success factors and limitations in establishing causality motivate a new theoretical approach. At least five criteria for this approach are evident. First, the theory should clearly identify one or more antecedents of design project success or failure. Second, the theory should be testable. Although this may seem obvious, rationalism has a strong and destructive tradition in design research [12]. Third, the theory should rest on a sound theoretical or empirical basis, derived from existing theory or observation. Fourth, the theory should produce actionable recommendations (e.g., 'model requirements using scenarios,' is more actionable than 'ensure executive support'). Fifth, the theory should possess reasonable face validity. The following section explores a potential theoretical basis – cognitive biases.

Table 1. Success, Abandonment and Risk Factors for Software Projects

#	Risk Factors [9]	Success Factors [4, 51, 52]	Abandonment Factors [3]
1	"Unrealistic or unarticulated project goals"	"User involvement"	"Unrealistic project goals and objectives"
2	"Inaccurate estimates of needed resources"	"Executive management support"	"Inappropriate project-team composition"
3	"Badly defined system requirements"	"Clear business objectives"	"Project management and control problems"
4	"Poor reporting of the project's status"	"Optimizing scope"	"Inadequate technical know-how"
5	"Unmanaged risks"	"Agile process"	"Problematic technology base/infrastructure"
6	"Poor communication among customers, developers, and users"	"Project manager expertise"	"Changing requirements"
7	"Use of immature technology"	"Financial management"	"Lack of executive support and commitment"
8	"Inability to handle the project's complexity"	"Skilled resources"	"Insufficient user commitment and involvement"
9	"Sloppy development practices"	"Formal methodology"	"Cost overruns and schedule delays"
10	"Poor project management"	"Standard tools and infrastructure"	
11	"Stakeholder politics"		
12	"Commercial pressures"		

3   Cognitive Biases

"Cognitive biases are cognitions or mental behaviors that prejudice decision quality in a significant number of decisions for a significant number of people" [13, p. 59]. For example, *default bias* [14] is the tendency to choose preselected options over superior, unselected options. Although dozens if not hundreds of cognitive biases have been identified, most research considers only one or a few examples [15]. Consequently, some biases seem to overlap. For example, the valence effect (the tendency to underestimate the probability of negative events and the inverse) is very similar to unrealistic optimism bias (the tendency to overestimate the likelihood of success). Given the proliferation of similar and interconnected biases, it may be useful to consider complexes of biases ("biasplexes") rather than individual biases.

To date, I have identified twelve biasplexes with potential ramifications for software development projects (Table 2). In the interest of space, however, this paper focuses on two biasplexes (Table 3) – one well-known and one new. Biasplexes may have both positive and negative consequences. For example, *unrealistic optimism* may encourage people to produce unrealistic project schedules. This is the *planning fallacy* [16], "the tendency to hold a confident belief that one's own project will proceed as planned, even while knowing that the vast majority of similar projects have run late" [17, p. 366]. The planning fallacy may lead to poor decisions and project abandonment [3]. However, *unrealistic optimism* may also help people to cope with stress [18].

Furthermore, biases (and biasplexes) may have at least four causes. First, heuristics "reduce the complex tasks of assessing probabilities and predicting values to simpler judgmental operations" [19, p. 1124]. For example, the anchoring and adjustment heuristic [cf. 20] estimates quantities by increasing or decreasing an existing estimate. This leads to anchoring bias, systematic insufficient adjustment from anchors among decision makers. Second, cognitive illusions – incorrect, unconscious inferences [cf., 21] – are not limited to the optical illusions that populate undergraduate psychology textbooks. The illusion of control [22] for instance, may lead to excessive optimism, i.e., optimism bias. Third, group processes may cause or exacerbate individual biases. For example, groupthink [23] may lead to collective excessive optimism by inhibiting critical thinking. Fourth, other individual psychological phenomena, here called principles, may further create or exacerbate biases. For example, the Pollyanna Principle [24] – the tendency to be overly positive in perception, memory and judgment – may reinforce optimism bias.

Debiasing is the process of inhibiting or removing the effects of biases [25, 26]. As many biases are highly robust (e.g., anchoring bias), debiasing may be extremely challenging [27]. Fischoff [25] suggests that where biases are caused by faulty "but perfectible" judges, they may be addressed by one of four increasingly potent interventions – 1) warn participants about biases in general; 2) identify particular biases in play and their magnitude; 3) provide feedback with personalized implications; 4) extended training. Subsequent empirical research has found that the first three of Fischoff's interventions often fail [cf. 28, 29], leaving the often expensive, sometimes impractical option of extended training. Fischoff however further argued that biases may be addressed by restructuring the person-task system. This strategy has been effective in software project management [e.g., 30] and other domains [e.g., 31].

Table 2. Twelve Biasplexes

Biasplex	Definition	Example Biases
Affective Forecasting	a collection of tendencies to overestimate the emotional impact of events	impact bias, hot-cold empathy gap
Causality Errors	a collection of tendencies to infer causal relationships without sufficient evidence	actor-observer bias, egocentric bias
Fixation	a collection of tendencies to disproportionately focus on one aspect of an event, object or situation, especially self-imposed or imaginary obstacles	design fixation, negativity bias
Framing Effects	"the tendency to give different responses to problems that have surface dissimilarities but that are really formally identical" [48, p. 88]	loss aversion, irrational escalation
Memory Errors	a collection of tendencies to remember ideas and events inaccurately	hindsight bias, self-serving bias
Miserly Information Processing	the tendency to avoid deep or complex information processing [48]	confirmation bias, belief bias
Overconfidence	the tendency to overestimate one's own skill, accuracy and control over one's self and environment	overconfidence effect, restraint bias
Perception Errors	a collection of tendencies to perceive situations inaccurately	selective perception, halo effect
Probabilistic Reasoning Errors	a collection of tendencies to err in numerical reasoning about chance and uncertainty.	base-rate neglect, subadditivity effect
Inertia	a collection of tendencies that increase disproportional preference for and defense of the status quo	see Table 3
Resistance to Self-criticism	a collection of tendencies to avoid recognizing one's mistakes or downplay their significance	outcome bias, myside bias
Unrealistic Optimism	the tendency to to make overly positive estimates and attributions	see Table 3

Table 3. Unrealistic Optimism and Inertia Biasplexes Expanded

Biasplex	Bias	Definition
Unrealistic Optimism	Wishful Thinking	the tendency to underestimate the likelihood of a negative outcome and vice versa [53]
	Valence Effect	the tendency to give undue weight to the degree to which an outcome is considered as positive or negative when estimating the probability of its occurrence [50]
	Optimism bias	the tendency for predicted outcomes to be more positive than actual outcomes [54]
	Normalcy Bias	systematically "underestimating the probability or extent of expected disruption" during a disaster [62, p. 273]
Inertia	Endowment Effect	"The tendency to demand much more to give up an object than one is willing to pay to acquire it" [55, p. 252]
	Belief Perseverance	"the tendency to maintain a belief even after the information that originally gave rise to it has been refuted or otherwise shown to be inaccurate" [56, p. 112]
	Anchoring Bias	"The tendency, in forming perceptions or making quantitative judgments of some entity under conditions of uncertainty, to give excessive weight to the initial starting value (or anchor), based on the first received information or one's initial judgment, and not to modify this anchor sufficiently in light of later information" [56, p. 51]
	Bandwagon Effect	"the tendency for large numbers of individuals, in social and sometimes political situations, to align themselves or their stated opinions with the majority opinion as they perceive it" (VandenBos 2007, p. 101)
	Semmelweis Reflex	unthinking rejection of new information that contradicts established beliefs or paradigms [61]
	Default Bias	the tendency to chose an pre-selected option regardless of its superiority or inferiority to other options [57]
	Mere Exposure Effect	"the ... increased liking for a stimulus that follows repeated, unreinforced exposure to that stimulus" [60, p. 231]
	Validity Effect	"the validity effect occurs when the mere repetition of information affects the perceived truthfulness of that information ... the validity effect occurs similarly for statements of fact that are true and false in origin, as well as political or opinion statements" [58, p. 211]
	System Justification	the tendency to defend and reinforce the existing social order [59]

Of course, using cognitive biases to inform information systems and design research is not new. Numerous studies have investigated cognitive biases in software engineering [e.g., 20, 32, 33], information modeling [e.g., 34], engineering design [e.g., 35], software project management [e.g., 36] and requirements determination [37]. In design science research specifically, [13] used cognitive biases to generate recommendations for decision support system development. However, a theory of the role of cognitive biases in software projects remains elusive.

4   Toward a New Theory

This section uses three scenarios to introduce a cognitive, variance theory of design project success. I begin with the scenario that inspired the theory.

Scenario 1. A seven-person software development team meets to estimate the time required to implement a variety of features for a virtual learning environment (VLE). The project manager describes a high-priority feature, a virtual gradebook. Robert, the team lead and most experienced developer, envisions a spreadsheet-like interface and estimates that it will take two days. Robert does not know that 1) the VLE needs to support both numerical and alphabetical grades, 2) different departments use different mappings, and 3) that the mapping criteria are not well documented. James, a junior developer who is aware of these complexities intuits that the two day estimate is insufficient and estimates that the feature may require three or four days, rather than two. The feature eventually takes three weeks to complete and forces a delay in the initial prototype, endangering the overall project.

   In this scenario, James used the anchoring and adjustment heuristic to estimate the feature development time. Anchoring bias caused a specific forecasting error, which negatively affected the success of the project. To avoid exactly this type of error, the

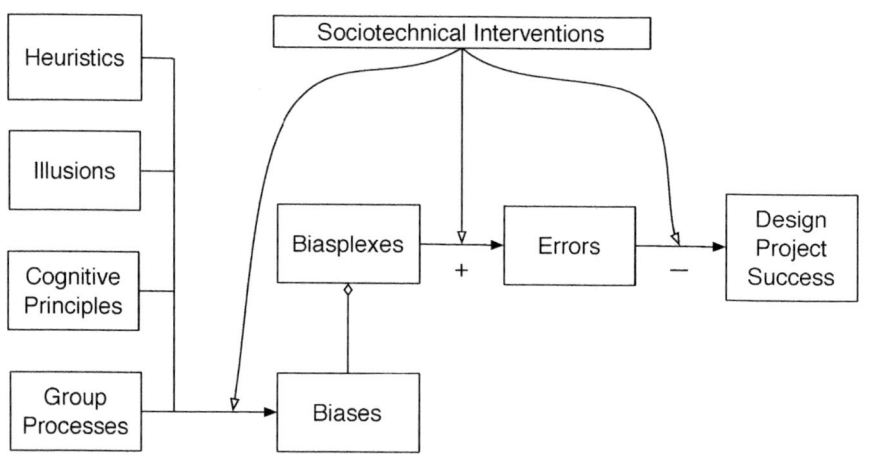

Fig. 1. A Variance Theory of Debiasing Software Project Participants

Note: Filled arrows indicate causation; unfilled arrows indicate moderating effects; plus and minus signs indicate direction of effect; unfilled diamond indicates aggregation.

software project management framework Scrum [38] includes *planning poker*, a practice where developers choose estimates in secret and show them simultaneously. Some empirical evidence indicates that "planning poker improve[s] the team's estimation performance in most cases" [39, p. 23] by preventing participants from anchoring on an initial estimate. This example motivated the proposed theory (Fig. 1, Table 4).

As described above, bias(plex)es are caused by heuristics, illusions, group processes and cognitive principles. As this relationship is squarely in the cognitive psychology domain, it is included only for completeness and is not the focus of the present research. In contrast, the relationship between biasplexes and errors and between errors and success are primary concerns to understand software development. Moreover, the moderating effect of sociotechnical interventions on these relationships may provide the basis for much design research. Scenarios two and three illustrate the use and usefulness of this framework.

Table 4. Theory Constructs Defined

Construct	Definition	Examples
Heuristic	a process or method that simplifies assessing probabilities or predicting values	availability, affect, default
Illusion	an incorrect, unconscious inference prompted by an external stimulus	illusion of control, illusory superiority
Cognitive Principle	a psychological disposition or tendency that interferes with optimal judgment	Pollyanna Principle, cognitive dissonance
Group Process	a shared psychological phenomenon that alters cognition or perception	Groupthink, firehouse effect
Bias	a predictable, systematic *error*	valance effect, default bias
Biasplex	a complex of similar, mutually-reinforcing biases	unrealistic optimism, inertia
Error	a deviation from optimal judgment	planning fallacy, sunk cost fallacy
Sociotechnical Intervention	a method, practice, technique, tool, program or concept that changes the behavior of one or more participants in a software development project	planning poker, checklists
Design Project Success	a multidimensional construct representing the achievement associated with a software design project from various stakeholder perspectives	developer learning, user satisfaction, client profits

Scenario 2. Following *Scenario 1*, suppose the same team has begun using planning poker to prevent anchoring bias in task estimation. This time, the project manager

asks for an estimate for integrating the VLE's gradebook with the university's legacy student information management system. Subconsciously, the developers are inherently optimistic; they assume that they can make minor changes to the legacy system (illusion of control), downplay the bureaucratic complexities of interacting with the legacy system's developers (optimism) and overestimate their own skill and productivity (Pollyanna Principle). These factors mutually reinforce each other, producing *Unrealistic Optimism*. The resulting error, an unrealistically optimistic estimate from each individual during planning poker, produces a cumulatively optimistic estimate, which threatens the project. While planning poker inhibited anchoring and adjustment, participants' shared optimism may still produce unrealistic schedules. While reference class forecasting [40, 41] – an effective technique for avoiding optimistic estimates – is readily available, a review of estimation methods in use was "not [...] able to identify a single study reporting a dominant use of formal estimation methods" [42, p. 228]. This raises the question, how can more effective forecasting methods gain traction in this domain?

Scenario 3. After circumventing some scheduling overruns, the VLE team is asked to deliver a document sharing feature. One of the developers points out that Microsoft SharePoint may be used for document sharing and that the university already has an unlimited license for it. The project manager asks if SharePoint supports collaborative editing, a desired aspect of document sharing. The developer responds with uncertainty but suggests worrying about advanced features like collaborative editing later while emphasizing how SharePoint is increasingly the default infrastructure choice within the university. The team then builds a SharePoint-based system, which delivers basic functionality in a familiar way but no more. This illustrates the combination of the default heuristic, bandwagon effect and status quo bias (the *Inertia* biasplex) producing a suboptimal decision. Alternatives including Google Apps, which is free and supports collaborative document editing innately, were not considered. This raises the question, what intervention could inhibit Inertia and stimulate better alternative evaluation?

In *Scenario 1*, the development team experienced a threat to success with a causal chain including the anchoring and adjustment heuristic, adjustment bias and a forecasting error. The planning poker intervention inhibits this kind of error and, being part of a popular project management framework, is commonly used in software development. *Scenario 2* is similar in that effective interventions have been identified but different in that model-based forecasting has not caught on. *Scenario 3* is also similar in that a biasplex manifests as a participant error; however, it differs in that no specific intervention has been developed to inhibit Inertia in this context. These scenarios illustrate the following four research streams implied by the proposed theory.

1. What errors are produced by biasplexes in software design projects?
2. How and to what extend do these errors affect Design Project Success?
3. What sociotechnical interventions can ameliorate the effects of biasplexes and resulting errors?
4. How can interventions be propagated through the software development industry?

These questions reveal the interdisciplinarity of the proposed research. Question one is behavioral research, which may be studied using both lab studies and ethnographic approaches. Question two is amenable to ethnography, case studies, surveys and econometric analysis of secondary datasets. Question three appears well suited to the design science approach [43, 44] and possibly action research [45]. Question four is especially challenging and may involve aspects of critical social theory, action research and social network analysis.

5 Conceptual Evaluation

Section Two identified five criteria on which to evaluate the proposed theory – clear antecedents, testability, sound basis, actionable recommendations and face validity. This section evaluates the theory with each criteria.

The proposed theory clearly identifies one class of design project success antecedents – participant errors – and reveals one related causal chain. Isolating errors in this way is motivated by the need to produce actionable recommendations. It does not preclude other causal chains.

The theory is clearly testable. The causes of cognitive biases are already a major subject of empirical study in psychology. Bias-induced errors may be both observed in case studies and demonstrated in lab studies. The effect of errors on specific problems and risks within a project is equally observable, at least through perceptual measures; measuring success however remains challenging.

Cognitive psychology provides ample theoretical basis for the proposed theory. Extensive literature on cognitive biases has developed over the past 40 years, with many quality books and papers on the subject [e.g., 21, 25, 46-48]. As cognitive biases affect everyone, mapping their established role into software development projects should be uncontroversial in principle.

The proposed theory produces two types of recommendations. The more general recommendation is to address common errors in software projects with sociotechnical interventions that modify the person-task system, inhibiting biases so that the errors do not manifest. More specific recommendations comprise the interventions themselves. For example, preliminary analysis suggests that Inertia may be inhibited by designating a Devil's Advocate during decision-making or drawing a high-level design tree showing previous decisions and alternatives. More generally, the theory is useful for stimulating hypotheses about particular errors. For example, one may hypothesize that Unrealistic Optimism will manifest as inflated expectations among clients and intended users, which would increase the chance of disappointment (cf. Expectation Confirmation Theory [49]).

Finally, this paper attempts to demonstrate face validity by using three scenarios (above) to illustrate the causal chain. The idea that software project participants fall prey to cognitive biases resulting in errors seems uncontroversial. As biases are systematic and predictable, it appears credible to hypothesize that the resulting errors will also be credible and predictable. The more difficult questions are the extent of success variance explained by these errors and what interventions may avoid them.

6 Conclusion

This paper's contribution is twofold. First, it introduces the term "biasplex" and identifies 12 such interconnected sets of cognitive biases. Second, it proposes a theory of debiasing software development projects including a chain of antecedents for software design project success and its relationship to sociotechnical interventions for debiasing participants. Conceptual evaluation of the theory appears promising and methods for empirical evaluation are suggested. Future research (in progress) entails hypothesizing common errors caused by each biasplex (Table 2), organized both by participant and by design activity, then identifying evidence of these errors in a series of case studies. This may be followed by developing interventions and testing their effectiveness in reducing the supported errors.

The proposed theory has several limitations. First, it focuses on a single antecedent of Design Project Success (but does not imply that this is the only or primary antecedent). Second, Design Project Success is not well understood, difficult to define and challenging to measure. Third, at this stage no characteristics of successful interventions can be stated other than noting that modest training on biases has been ineffective while extensive, rapid, personalized feedback or changing the person-task system has been effective in previous studies. Finally, as empirical evaluation is still ongoing, the theory is evaluated conceptually. While conceptual evaluation is inherently limited, it is appropriate for this type of greenfield research-in-progress.

References

1. Desmond, J.P.: System Integrators, Outsourcers Gain Again as IT Guards in-House Resources. Software Magazine, King Content Co. (2010),
   http://www.softwaremag.com/focus-areas/the-software-500-industry/featured-articles/system-integrators-outsourcers-gain-again-as-it-guards-in-house-resources/
2. Gartner: Gartner Says Worldwide IT Spending to Grow 5.3 Percent in 2010. Gartner, Inc. (2010), http://www.gartner.com/it/page.jsp?id=1339013
3. Ewusi-Mensah, K.: Software Development Failures. MIT Press (2003)
4. Standish Group: Chaos Summary 2009, Boston, MA, USA (2009)
5. Moløkken-Østvold, K.: A Review of Surveys on Software Effort Estimation. In: Proceedings of International Symposium on Empirical Software Engineering (ISESE 2003), pp. 223–230. IEEE, Rome (2003)
6. Ralph, P., Wand, Y.: A Proposal for a Formal Definition of the Design Concept. In: Lyytinen, K., Loucopoulos, P., Mylopoulos, J., Robinson, B. (eds.) Design Requirements Engineering. LNBIP, vol. 14, pp. 103–136. Springer, Heidelberg (2009)
7. Van de Ven, A.H.: Engaged Scholarship: A Guide for Organizational and Social Research. Oxford University Press, Oxford (2007)
8. Celdman, J.: Designing with Web Standards. New Riders, Berkely (2007)
9. Charette, R.N.: Why Software Fails. IEEE Spectrum Online (2005),
   http://www.spectrum.ieee.org/sep05/1685
10. Koppel, R., Metlay, J.P., Cohen, A., Abaluck, B., Localio, A.R., Kimmel, S.E., Strom, B.L.: Role of Computerized Physician Order Entry Systems in Facilitating Medication Errors. Journal of the American Medical Association 293, 1197–1203 (2005)

11. Keizer, G.: A Third of New Pcs Being Downgraded to XP, Says Metrics Researcher. Computerworld (2008)
12. Ralph, P.: Introducing an Empirical Model of Design. In: Proceedings of The 6th Mediterranean Conference on Information Systems, Limassol, Cyprus (2011)
13. Arnott, D.: Cognitive Biases and Decision Support Systems Development: A Design Science Approach. Information Systems Journal 16, 55–78 (2006)
14. Samuelson, W., Zeckhauser, R.J.: Status Quo Bias in Decision Making. Journal of Risk and Uncertainty 1, 7–59 (1988)
15. Gheorghiu, V., Molz, G., Pohl, R.: Suggestion and Illusion. In: Pohl, R. (ed.) Cognitive Illusions: A Handbook on Fallacies and Biases in Thinking, Judgement and Memory, pp. 399–421. Psychology Press, Hove (2004)
16. Kahneman, D., Tversky, A.: Prospect Theory: An Analysis of Decision under Risk. Econometrica 47, 263–291 (1979)
17. Buehler, R., Griffin, D., Ross, M.: Exploring the "Planning Fallacy": Why People Underestimate Their Task Completion Times. Journal of Personality and Social Psychology 67, 366–381 (1994)
18. Taylor, S.E., Kemeny, M.E., Aspinwall, L.G., Schneider, S.G., Rodriguez, R., Herbert, M.: Optimism, Coping, Psychological Distress, and High-Risk Sexual Behavior among Men at Risk for Acquired Immunodeficiency Syndrome (Aids). Journal of Personality and Social Psychology 63, 460–473 (1992)
19. Tversky, A., Kahneman, D.: Judgment under Uncertainty: Heuristics and Biases. Science 185, 1124–1131 (1974)
20. Parsons, J., Saunders, C.: Cognitive Heuristics in Software Engineering: Applying and Extending Anchoring and Adjustment to Artifact Reuse. IEEE Transactions on Software Engineering 30, 873–888 (2004)
21. Pohl, R. (ed.): Cognitive Illusions. Psychology Press, East Sussex (2004)
22. Langer, E.: The Illusion of Control. Journal of Personality and Social Psychology 32, 311–328 (1975)
23. Janis, I.L.: Groupthink: Psychological Studies of Policy Decisions and Fiascoes. Houghton Mifflin (1982)
24. Matlin, M.: Pollyanna Principle. In: Pohl, R.F. (ed.) Cognitive Illusions: A Handbook on Fallacies and Biases in Thinking, Judgement and Memory, pp. 255–272. Psychology Press, Hove (2004)
25. Fischoff, B.: Debiasing. In: Kahneman, D., Slovic, P., Tversky, A. (eds.) Judgment under Uncertainty: Heuristics and Biases. Cambridge University Press, Cambridge (1982)
26. Larrick, R.P.: Debiasing. In: Blackwell Handbook of Judgment and Decision Making, pp. 316–338. Blackwell Publishing Ltd. (2008)
27. Wilson, T., Centerbar, D., Brekke, N.: Mental Contamination and the Debiasing Problem. In: Gilovich, T., Griffin, D., Kahneman, D. (eds.) Heuristics and Biases: The Psychology of Intuitive Judgment, pp. 185–200. Cambridge University Press, Cambridge (2002)
28. Pronin, E., Lin, D.Y., Ross, L.: The Bias Blind Spot: Perceptions of Bias in Self Versus Others. Personality and Social Psychology Bulletin 28, 369–381 (2002)
29. Pronin, E.: Perception and Misperception of Bias in Human Judgment. Trends in Cognitive Sciences 11, 37–43 (2006)
30. Jørgensen, M.: Identification of More Risks Can Lead to Increased Over-Optimism of and Over-Confidence in Software Development Effort Estimates. Information & Software Technology 52, 506–516 (2009)
31. Thaler, R., Benartzi, S.: Save More Tomorrow: Using Behavioral Economics to Increase Employee Saving. The Journal of Political Economy 112, S164–S187 (2004)

32. Leventhal, L., Teasley, B., Rohlman, D., Instone, K.: Positive Test Bias in Software Testing among Professionals: A Review. In: Bass, L.J., Unger, C., Gornostaev, J. (eds.) EWHCI 1993. LNCS, vol. 753, pp. 210–218. Springer, Heidelberg (1993)
33. Stacy, W., MacMillan, J.: Cognitive Bias in Software Engineering. Communications of ACM 38, 57–63 (1995)
34. Siau, K., Wand, Y., Benbasat, I.: When Parents Need Not Have Children — Cognitive Biases in Information Modeling. In: Constantopoulos, P., Vassiliou, Y., Mylopoulos, J. (eds.) CAiSE 1996. LNCS, vol. 1080, pp. 402–420. Springer, Heidelberg (1996)
35. Busby, J., Payne, K.: The Situated Nature of Judgement in Engineering Design Planning. Journal of Engineering Design 9, 271–291 (1998)
36. Snow, A.P., Keil, M., Wallace, L.: The Effects of Optimistic and Pessimistic Biasing on Software Project Status Reporting. Information & Management 44, 130–141 (2007)
37. Browne, G.J., Ramesh, V.: Improving Information Requirements Determination: A Cognitive Perspective. Information & Management 39, 625–645 (2002)
38. Schwaber, K.: Agile Project Management with Scrum. Microsoft Press (2004)
39. Nils, C.H.: An Empirical Study of Using Planning Poker for User Story Estimation. In: Proceedings of Agile 2006, pp. 23–34. IEEE Computer Society, Visegrád (2006)
40. Flyvbjerg, B.: From Nobel Prize to Project Management: Getting Risks Right. Project Management Journal 37, 5 (2006)
41. Lovallo, D., Kahneman, D.: Delusions of Success. Harvard Business Review 81, 56–63 (2003)
42. Molokken, K., Jorgensen, M.: A Review of Software Surveys on Software Effort Estimation. In: Proceedings of International Symposium on Empirical Software Engineering, pp. 223–230. ACM-IEEE, Rome, Italy (2003)
43. Hevner, A.R., March, S.T., Park, J., Ram, S.: Design Science in Information Systems Research. MIS Quarterly 28, 75–105 (2004)
44. Hevner, A., Chatterjee, S.: Design Research in Information Systems: Theory and Practice. Springer, Heidelberg (2010)
45. Sein, M., Henfridsson, O., Purao, S., Rossi, M., Lindgren, R.: Action Design Research. MIS Quarterly 35 (2011)
46. Gilovich, T., Griffin, D., Kahneman, D. (eds.): Heuristics and Biases: The Psychology of Intuitive Judgment. Cambridge University Press, Cambridge (2002)
47. Newell, A., Simon, H.: Human Problem Solving. Prentice-Hall, Inc. (1972)
48. Stanovich, K.: What Intelligence Tests Miss: The Psychology of Rational Thought. Yale University Press, New Haven (2009)
49. Oliver, R.L.: A Cognitive Model of the Antecedents and Consequences of Satisfaction Decisions. Journal of Marketing Research 17, 460–469 (1980)
50. Taylor, N.: Making Actuaries Less Human: Lessons from Behavioural Finance. The Staple Inn Actuarial Society Meeting (2000)
51. Standish Group: Chaos Database: Chaos Surveys Conducted from 1994 to Fall 2004 (2006)
52. Hartmann, D.: Interview: Jim Johnson of the Standish Group. InfoQueue (2006)
53. Poses, R.M., Anthony, M.: Availability, Wishful Thinking, and Physicians' Diagnostic Judgments for Patients with Suspected Bacteremia. Medical Decision Making 11, 159–168 (1991)
54. Armor, D., Taylor, S.: When Predictions Fail: The Dilemma of Unrealistic Optimism. In: Gilovich, T., Griffin, D., Kahneman, D. (eds.) Heuristics and Biases: The Psychology of Intuitive Judgment. Cambridge University Press, Cambridge (2002)
55. Colman, A.: A Dictionary of Psychology. Oxford University Press, Oxford (2009)

56. VandenBos, G. (ed.): APA Dictionary of Psychology. American Psychological Association, Washington, DC, USA (2007)
57. Todd, P.M., Gigerenzer, G.: Environments That Make Us Smart: Ecological Rationality. Current Directions in Psychological Science 16, 167–171 (2007)
58. Renner, C.H.: Validity Effect. In: Pohl, R.F. (ed.) Cognitive Illusions: A Handbook on Fallacies and Biases in Thinking, Judgement and Memory, pp. 201–213. Psychology Press, Hove (2004)
59. Jost, J.T., Banaji, M.R.: The Role of Stereotyping in System-Justification and the Production of False Consciousness. British Journal of Social Psychology 33, 1–27 (1994)
60. Bornstein, R., Carver-Lemley, C.: Mere Exposure Effect. In: Pohl, R.F. (ed.) Cognitive Illusions: A Handbook on Fallacies and Biases in Thinking, Judgement and Memory, pp. 215–234. Psychology Press, Hove (2004)
61. Shore, B.: Systematic Biases and Culture in Project Failures. Project Management Journal 39, 5–16 (2008)
62. Omer, H., Alon, N.: The Continuity Principle: A Unified Approach to Disaster and Trauma. American Journal of Community Psychology 22, 273–287 (1994)

OBCAS - An Ontology-Based Cluster Analysis System

Janusz Tuchowski[1], Katarzyna Wójcik[1], Paweł Lula[1],
and Grażyna Paliwoda-Pękosz[2]

[1] Department of Computational Systems
[2] Department of Computer Science
Cracow University of Economics, Rakowicka 27, 31-510 Krakow, Poland
{janusz.tuchowski,katarzyna.wojcik,pawel.lula,
grazyna.paliwoda-pekosz}@uek.krakow.pl

Abstract. The main objective of the paper is to present the OBCAS - an ontology-based cluster analysis system that was implemented in Java. The system takes as an input the data described by an ontology, allows the user to choose individuals, and produces the similarity matrix. The core component of this system is a module that counts three kinds of similarity measures: taxonomy, relationship and attribute. The current outcome of the system is an aggregate similarity matrix that can be further processed by any statistical package.

Keywords: similarity measures, ontology, cluster analysis, framework implementation.

1 Introduction

The growing popularity of using ontologies to enrich the data description and capturing the reality calls for system that can help in the analysis of these data. One of the basic techniques of data analysis is the cluster analysis [2]. The crucial role in all cluster analysis algorithms plays counting the similarity between individuals. The exhaustive presentation of similarity measures that can be used in the ontology context can be found in [1]. Though in research literature there are some papers that focus on clustering of ontology-based data (e.g. [4]) still there are not many systems that can perform this analysis/clustering. Besides, they are usually based on ontologies that represent knowledge from a certain domain, e.g. clustering of biomedical data described by the Gene Ontology [6] or geospatial data described by GeoCo ontology [5]. This was the main motivation for the design and implementation of the universal system that is presented in this paper. It implements the framework that was introduced by Lula&Paliwoda-Pękosz [3].

This paper will take the following structure: the framework of ontology-based cluster analysis and the system schema will be presented in section 2, section 3 will tackle some implementation issues and will demonstrate the main functionality of the system, and in section 4 the results of experiments that were performed using this system will be discussed. Finally, section 5 will outline the possible paths for the system development.

S. Wrycza (Ed.): SIGSAND/PLAIS 2011, LNBIP 93, pp. 106–112, 2011.

2 Design of Ontology-Based Cluster Analysis System (OBCAS)

An Ontology-Based Cluster Analysis Framework (OBCAF) is based on the assumption that the similarity measure between two instances of an ontology (I_i, I_j) is a function of taxonomy (TS), relationship (RS), and attribute similarities (AS):

$$sim(I_i, I_j) = f_{agr}\left(TS(I_i, I_j), RS(I_i, I_j), AS(I_i, I_j)\right) \tag{1}$$

It generalizes the model proposed by Maedche&Zacharias [4] in which as an aggregation function of the three kinds of similarities mentioned above a weighted average was adopted.

Taxonomy similarity is counted between classes to which compared objects belong; it depends of the hierarchy of classes. Relationship similarity compares the links to other objects described by an ontology; it depends of the neighborhood of objects according to relationships by which they are connected. Finally, attribute similarity reflects the similarity between attribute values; it can be counted in different ways in accordance with types of these attributes (numerical values, intervals, nominal values, strings, texts, sets, lists). The detailed description of various ways of counting these similarity measures can be found in [3].

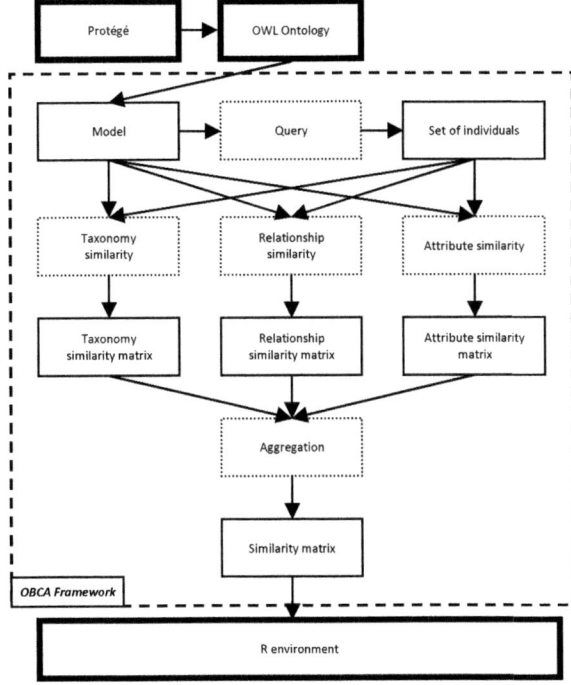

Fig. 1. The Ontology-Based Cluster Analysis System

On the basis of OBCAF the system for performing cluster analysis of ontology-based data was proposed [3] (Fig. 1). It takes as an input ontology in OWL that is designed in an external tool. e.g. Protégé[1]. This ontology is represented in the system by a *Model*. The *Query* module allows to choose individuals for comparison (*Set of individuals*). In the next stage three kinds of similarities are counted (taxonomy, relationship, and attribute). As a result three square matrixes of similarities are produced. They are aggregated and finally the aggregated similarity matrix is obtained. This matrix can be further processed by statistical external tools, e.g. package R^2.

The current implementation of the system is presented in the next section.

3   Implementation of OBCAS

The OBCAS was implemented in Java. It uses the Jena Ontology API[3] package to access ontology data and SimPack[4] Java library for counting attribute similarity for different types of attributes. Both those tools were imported as libraries into Java program. When ontology objects or relations were referred to, Jena defined functions were used. Some of measures implemented in SimPack were used for the attribute similarity counting.

OBCAS takes as an input an ontology in OWL with objects (individuals) described by that ontology, counts three kinds of similarities, and aggregates them. Currently, the aggregation function is the weighted average.

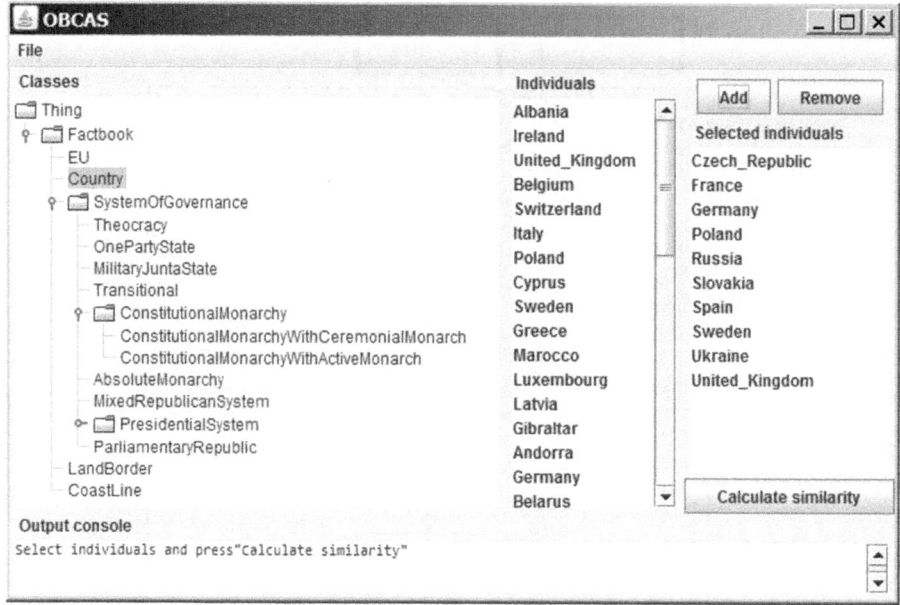

Fig. 2. The graphic user interface of OBCAS

---

[1] http://protege.stanford.edu/
[2] http://www.r-project.org/
[3] http://jena.sourceforge.net/ontology/
[4] http://www.ifi.uzh.ch/ddis/simpack.html

Using a graphic interface (Fig. 2) user can choose an ontology (*File* menu). The classes of the ontology will be displayed in the left part of the window (*Classes*). In the middle of the window, the individuals of the selected class are displayed (*Individuals*). User can add individuals to the right window (*Selected individuals*) and start calculation by pressing the button *Calculate similarity*. In the bottom window the results will appear: four similarity matrixes (taxonomy, relationship, attribute, and aggregated). Besides, the output data are saved in a CSV file.

Additionally, in the configuration file user can set different parameters for calculations, e.g.: the similarity for missing attributes values (maximal – 1 or minimal – 0), the similarity of individuals without attributes (maximal – 1 or minimal – 0), and the method for the string comparison.

4   Evaluation of OBCAS

4.1   Data and Methodology

The main functions of the OBCAS will be demonstrated on an example ontology that describes countries. This ontology was developed on the basis of the World Factbook[5]. All data were as of 2010 except for the population data that were as of July 2011. The class hierarchy of this ontology is presented in Fig. 2 (the left window). Each country was described by a set of attributes: *TotalImportValue*, *TotalExportValue*, *GDPPerCapita*, *CountryPopulation*, *CountryArea*, *CountryName*, *CountryCapital*. The relationships between classes were depicted in Fig. 3. The relationship similarity of two countries (individuals of the Country class) is greater for countries that have more common borders represented by instances of the class LandBorder, similar length of coastline represented by instances of the class CoastLine, and similar system of governance.

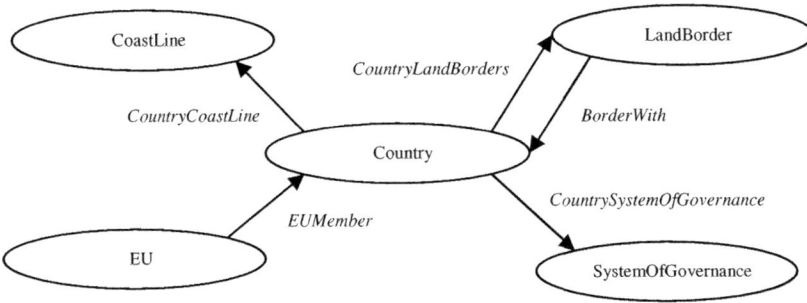

Fig. 3. The relationships between classes in the Factbook ontology

In order to demonstrate system performance 10 countries were chosen (Fig. 2, the right window). The parameters in the configuration file were set as following:

---

[5] https://www.cia.gov/library/publications/the-world-factbook/

- the similarity for missing attributes values: maximal (1),
- the similarity of instances without attributes: maximal (1).
- the similarity of string attributes: Levenshtein metrics.

4.2 Results

The taxonomy similarity for all pairs of countries is 1 because they belong to the same class. Tables 1-2 present similarity matrixes for attribute and relationship similarities. Table 3 depicts aggregated similarity matrix.

Table 1. Attribute similarity matrix

Country	Czech_Republic	France	Germany	Poland	Russia	Slovakia	Spain	Sweden	Ukraine	United_Kingdom
Czech_Republic	1	0,61	0,51	0,65	0,36	0,79	0,60	0,64	0,58	0,48
France	0,61	1	0,57	0,53	0,34	0,54	0,67	0,48	0,39	0,70
Germany	0,51	0,57	1	0,55	0,31	0,50	0,52	0,50	0,41	0,52
Poland	0,65	0,53	0,55	1	0,47	0,68	0,67	0,62	0,62	0,52
Russia	0,36	0,34	0,31	0,47	1	0,36	0,41	0,32	0,37	0,37
Slovakia	0,79	0,54	0,50	0,68	0,36	1	0,57	0,60	0,64	0,44
Spain	0,60	0,67	0,52	0,67	0,41	0,57	1	0,68	0,53	0,64
Sweden	0,64	0,48	0,50	0,62	0,32	0,60	0,68	1	0,50	0,58
Ukraine	0,58	0,39	0,41	0,62	0,37	0,64	0,53	0,50	1	0,41
United_Kingdom	0,48	0,70	0,52	0,52	0,37	0,44	0,64	0,58	0,41	1

Table 2. Relationship similarity matrix

Country	Czech_Republic	France	Germany	Poland	Russia	Slovakia	Spain	Sweden	Ukraine	United_Kingdom
Czech_Republic	1	0,87	0,92	0,92	0,87	1,00	0,87	0,87	0,91	0,85
France	0,87	1	0,88	0,87	0,92	0,87	0,88	0,87	0,87	0,84
Germany	0,92	0,88	1	0,92	0,87	0,91	0,87	0,87	0,92	0,86
Poland	0,92	0,87	0,92	1	0,88	0,92	0,87	0,87	0,92	0,85
Russia	0,87	0,92	0,87	0,88	1	0,87	0,87	0,87	0,88	0,86
Slovakia	1,00	0,87	0,91	0,92	0,87	1	0,87	0,87	0,92	0,85
Spain	0,87	0,88	0,87	0,87	0,87	0,87	1	0,92	0,87	0,87
Sweden	0,87	0,87	0,87	0,87	0,87	0,87	0,92	1	0,87	0,77
Ukraine	0,91	0,87	0,92	0,92	0,88	0,92	0,87	0,87	1	0,86
United_Kingdom	0,85	0,84	0,86	0,85	0,86	0,85	0,87	0,77	0,86	1

Table 3. Aggregated similarity matrix

Country	Czech_Republic	France	Germany	Poland	Russia	Slovakia	Spain	Sweden	Ukraine	United_Kingdom
Czech_Republic	1	0,83	0,81	0,85	0,74	0,93	0,82	0,84	0,83	0,78
France	0,83	1	0,81	0,80	0,75	0,80	0,85	0,79	0,76	0,85
Germany	0,81	0,81	1	0,82	0,73	0,80	0,80	0,79	0,77	0,79
Poland	0,85	0,80	0,82	1	0,78	0,87	0,85	0,83	0,85	0,79
Russia	0,74	0,75	0,73	0,78	1	0,75	0,76	0,73	0,75	0,74
Slovakia	0,93	0,80	0,80	0,87	0,75	1	0,81	0,82	0,85	0,76
Spain	0,82	0,85	0,80	0,85	0,76	0,81	1	0,87	0,80	0,84
Sweden	0,84	0,79	0,79	0,83	0,73	0,82	0,87	1	0,79	0,78
Ukraine	0,83	0,76	0,77	0,85	0,75	0,85	0,80	0,79	1	0,76
United_Kingdom	0,78	0,85	0,79	0,79	0,74	0,76	0,84	0,78	0,76	1

The aggregate similarity matrix was transferred to the R-environment where the clustering process was performed with the use of the Ward method. The outcome (dendrogram) of this clustering was presented in Fig. 4.

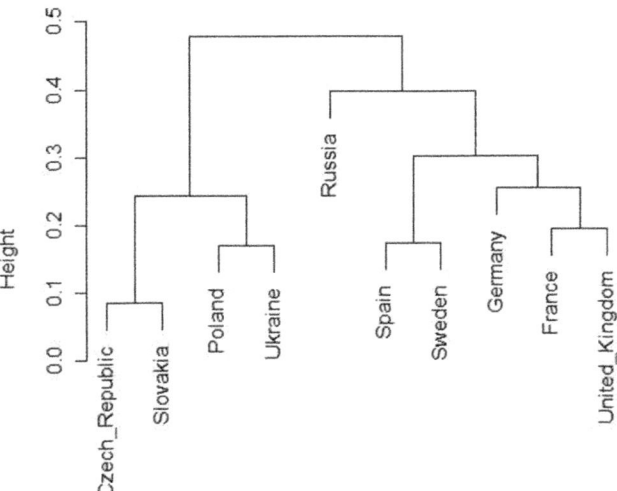

Fig. 4. Cluster dendrogram.

4.3 Discussion of Results

The preliminary results of the system performance show that the system can be used for an analysis of any domain that is described by an ontology. The system implementation is not linked to any specific ontology. This approach allows also to test different kinds of similarity measures and can be useful in an analysis of complex structures. In addition, it can be easily expanded thus in the future more features can be added. Particularly, a user is not limited to measures proposed in SimPack; other similarity measures can be easily added to the application. Besides, the way of saving results is convenient for users-the output file can be read in many different external tools. Finally, OBCAS's graphical interface makes the system easy to use.

5 Conclusion

In the paper the Ontology-Based Cluster Analysis System was presented. It allows to test a performance of a variety of similarity measures. The preliminary results of testing the system are encouraging but there is still room for improvement. The system can be further developed in the following direction: a fuller integration of the system with R-environment, incorporation of other kinds of similarity measures for counting string similarity, e.g. similarity measures based on Wordnet, Web, and Wikipedia, adding the possibility of choosing a subset of class attributes that can be taken into account during calculation of attribute similarity, and finally allowing the user to choose different aggregation function; currently system allows only to count weighted average of taxonomy, relationship, and attribute similarities.

References

1. Euzenat, J., Shvaiko, P.: Ontology Matching. Springer, Heidelberg (2007)
2. Han, J., Kamber, M.: Data Mining: Concepts and Techniques, 2nd edn. Morgan Kaufmann (2006)
3. Lula, P., Paliwoda-Pękosz, P.: An Ontology-based Cluster Analysis Framework. In: Duke, A., Hepp, M., Bontcheva, K., Vilain, M. (eds.) Proceedings of the First International Workshop on Ontology-Supported Business Intelligence (OBI 2008), vol. 308. ACM Press (2008), http://portal.acm.org/citation.cfm?id=1452574 ISBN 978-1-60558-219-1
4. Maedche, A., Zacharias, V.: Clustering Ontology-Based Metadata in the Semantic Web. In: Elomaa, T., Mannila, H., Toivonen, H. (eds.) PKDD 2002. LNCS (LNAI), vol. 2431, pp. 348–360. Springer, Heidelberg (2002)
5. Wang, X., Gu, W., Ziébelin, D., Hamilton, H.: An Ontology-Based Framework for Geospatial Clustering. International Journal of Geographical Information Science 24(1), 1601–1630 (2010)
6. Wolting, C., McGlade, J., Tritchler, D.: Cluster analysis of protein array results via similarity of Gene Ontology annotation. BMC Bioinformatics 7, 338 (2006)

Author Index

GPSR Compliance

*The European Union's (EU) General Product Safety Regulation (GPSR)
is a set of rules that requires consumer products to be safe and our
obligations to ensure this.*

*If you have any concerns about our products, you can contact us on
ProductSafety@springernature.com*

In case Publisher is established outside the EU, the EU authorized
representative is:

Springer Nature Customer Service Center GmbH
Europaplatz 3
69115 Heidelberg, Germany

Batch number: 09490872

Printed by Printforce, the Netherlands